Understanding Conflicts about Wildlife

Studies of the Biosocial Society

General Editor: **Catherine Panter-Brick,** Professor of Anthropology, Yale University

The Biosocial Society is an international academic society engaged in fostering understanding of human biological and social diversity. It draws its membership from a wide range of academic disciplines, particularly those engaged in 'boundary disciplines' at the intersection between the natural and social sciences, such as biocultural anthropology, medical sociology, demography, social medicine, the history of science and bioethics. The aim of this series is to promote interdisciplinary research on how biology and society interact to shape human experience and to serve as advanced texts for undergraduate and postgraduate students.

Volume 1
Race, Ethnicity, and Nation: Perspectives from Kinship and Genetics
Edited by Peter Wade

Volume 2
Health, Risk, and Adversity
Edited by Catherine Panter-Brick and Agustín Fuentes

Volume 3
Substitute Parents: Biological and Social Perspectives on Alloparenting in Human Societies
Edited by Gillian Bentley and Ruth Mace

Volume 4
Centralizing Fieldwork: Critical Perspectives from Primatology, Biological and Social Anthropology
Edited by Jeremy MacClancy and Agustín Fuentes

Volume 5
Human Diet and Nutrition in Biocultural Perspective: Past Meets Present
Edited by Tina Moffat and Tracy Prowse

Volume 6
Identity Politics and the New Genetics: Re/Creating Categories of Difference and Belonging
Edited by Katharina Schramm, David Skinner and Richard Rottenburg

Volume 7
Ethics in the Field: Contemporary Challenges
Edited by Jeremy MacClancy and Agustín Fuentes

Volume 8
Health and Difference: Rendering Human Engagement in Colonial Engagements
Edited by Alexandra Widmer and Veronika Lipphardt

Volume 9
Understanding Conflicts about Wildlife: A Biosocial Approach
Edited by Catherine M. Hill, Amanda D. Webber and Nancy E.C. Priston

Understanding Conflicts about Wildlife

A Biosocial Approach

Edited by
Catherine M. Hill, Amanda D. Webber
and Nancy E.C. Priston

berghahn
NEW YORK · OXFORD
www.berghahnbooks.com

First published in 2017 by
Berghahn Books
www.berghahnbooks.com

©2017 Catherine M. Hill, Amanda D. Webber and Nancy E.C. Priston

All rights reserved. Except for the quotation of short passages for the purposes of criticism and review, no part of this book may be reproduced in any form or by any means, electronic or mechanical, including photocopying, recording, or any information storage and retrieval system now known or to be invented, without written permission of the publisher.

Library of Congress Cataloging-in-Publication Data

Names: Hill, Catherine M., editor. | Webber, Amanda D., editor. | Priston, Nancy E.C., editor.
Title: Understanding conflicts about wildlife : a biosocial approach / edited by Catherine M. Hill, Amanda D. Webber, and Nancy E.C. Priston.
Description: New York : Berghahn Books, 2017. | Series: Studies of the biosocial society ; volume 9 | Includes bibliographical references and index.
Identifiers: LCCN 2016054879 (print) | LCCN 2016056906 (ebook) | ISBN 9781785334627 (hardback : alk. paper) | ISBN 9781785334634 (eBook)
Subjects: LCSH: Human-animal relationships. | Wildlife management. | Wildlife conservation. | Wildlife depredation.
Classification: LCC QL85 .U525 2017 (print) | LCC QL85 (ebook) | DDC 333.95/4--dc23
LC record available at https://lccn.loc.gov/2016054879

British Library Cataloguing in Publication Data

A catalogue record for this book is available from the British Library

ISBN 978-1-78533-462-7 (hardback)
ISBN 978-1-78533-463-4 (ebook)

Contents

List of Figures and Tables — vii

Acknowledgements — ix

Introduction. Complex Problems: Using a Biosocial Approach to Understanding Human-Wildlife Interactions — 1
Catherine M. Hill

1. People, Perceptions and 'Pests': Human-Wildlife Interactions and the Politics of Conflict — 15
Phyllis C. Lee

2. Block, Push or Pull? Three Responses to Monkey Crop-Raiding in Japan — 36
John Knight

3. Unintended Consequences in Conservation: How Conflict Mitigation May Raise the Conflict Level—The Case of Wolf Management in Norway — 49
Ketil Skogen

4. Badger-Human Conflict: An Overlooked Historical Context for Bovine TB Debates in the UK — 65
Angela Cassidy

5. Savage Values: Conservation and Personhood in Southern Suriname — 95
Marc Brightman

6. Wildlife Value Orientations as an Approach to Understanding the Social Context of Human-Wildlife Conflict — 107
Alia M. Dietsch, Michael J. Manfredo and Tara L. Teel

7. **A Long-Term Comparison of Local Perceptions of Crop Loss to**   127
   Wildlife at Kibale National Park, Uganda: Exploring Consistency
   Across Individuals and Sites
   Lisa Naughton-Treves, Jessica L'Roe, Andrew L'Roe and Adrian Treves

8. **Conservation Conflict Transformation: Addressing the Missing**   148
   Link in Wildlife Conservation
   Francine Madden and Brian McQuinn

9. **Engaging Farmers and Understanding Their Behaviour to Develop**   170
   Effective Deterrents to Crop Damage by Wildlife
   Graham E. Wallace and Catherine M. Hill

10. **Using Geographic Information Systems at Sites of Negative**   194
    Human-Wildlife Interactions: Current Applications and
    Future Developments
    Amanda D. Webber, Stewart Thompson, Neil Bailey and Nancy E.C. Priston

Index   211

List of Figures and Tables

Figures

1.1. Circles of hierarchical and interacting contexts underlying and determining people's perception of 'conflict' with wildlife. Economic, social and political power are indicated by the size of the circle, while the flow of financial incentives and benefits (indicated by the arrows) tends to be outwards from the smaller circles to the national, leaving local people to share only the risks and costs. At each level, the media – local or global – can exploit issues for their own financial or political gain.   17

1.2. The concept of the socio-zoological scale: adapted and modified from Arluke and Sanders (1996) and Costa (2011). The simple hierarchical structure represents an attempt to access beliefs about different species based on qualities and values attributed to these species by people, and assumes that there is a relatively deterministic categorization (good/bad) in operation to organize these beliefs.   21

3.1. Wolf packs (circles) and territorial pairs (triangles) in Norway and Sweden during winter 2015–2016. The Norwegian management zone for wolves is indicated (hatched).   53

3.2. Management regions for large carnivores in Norway.   56

4.1. Anti-cull graffiti, Leamington Spa, United Kingdom, July 2015.   67

4.2. Traditional coat of arms and modern corporate logo of the UK borough of Broxbourne, located in the southeast of the country.   72

4.3. Logo of the UK Wildlife Trusts.   73

4.4. 'Tommy Brock', illustration from Beatrix Potter's *The Tale of Mr. Tod* (1912: 21).   74

4.5. 'Bring It'; *Daily Mash* badger T-shirt design.   85

viii   List of Figures and Tables

6.1. Percentage of Mutualists by income across states from a 2004 survey of residents living in the western United States. 115

6.2. Percentage of Mutualists by education across states from a 2004 survey of residents living in the western United States. 116

6.3. Percentage of Mutualists by urbanization across states from a 2004 survey of residents living in the western United States. 116

6.4. Distribution of two wildlife value orientation types (i.e. Mutualists or Utilitarians) by county from a 2009 survey of Washington residents. 118

6.5. Percentage of residents by county accepting of wolves being moved from one area to another to help establish new wolf populations from a 2009 survey of Washington residents. 120

7.1. Study sites at Kibale National Park, Uganda. 130

8.1. 'Levels of Conflict' indicating the three levels of conflict that may exist in any conflict situation. 154

8.2. Conflict Intervention Triangle Model. 158

9.1. Map showing the location of Budongo Forest Reserve in northwestern Uganda. 175

Tables

1.1. Synopsis of sociocultural factors affecting attitudes from two studies in East Africa. 20

1.2. Comparison of livelihood risk analysis for people in conflict contexts; risk mapping based on participatory interviews (e.g. Smith, Barrett and Box 2000) or weighted rank index of household reports. Risk index values range between 0 and 1, with 1 being indicative of a highly ranked risk. 23

1.3. Perceptions of pests in East African areas with primates plus elephants present and where similar methods for assessing pests were used. Respondents were asked to rank what species they thought their top crop pests were, and a summed rank index was derived from these rankings (see Gillingham 1998). Rank 1 = most highly ranked, NP = not present to be ranked, blank = not mentioned. 25

6.1. Examples of correlation coefficients examining the relationship between wildlife value orientations and attitudinal responses to wildlife issues from a 2004 survey of western US residents. 111

6.2. Comparison of wildlife value orientation types on acceptability of providing more recreational hunting opportunities in response to black bears that are getting into trash and pet-food containers from a 2004 survey of residents living in the western United States.   113

7.1. Individual-, farm- and village-level predictors of reported crop loss to four wildlife species around Kibale National Park, Uganda: 1994 and 2011–2012.   136

9.1. Deterrents trialled, their capacity to reduce crop damage by animals, animal species deterred and farmers' evaluations.   182

Acknowledgements

This book is the result of a workshop held by the Anthropology Centre for Conservation, the Environment and Development (ACCEND) at Oxford Brookes University on 16–17 September 2011. We are very grateful to our various supporters: the department of Social Sciences, Oxford Brookes University; the Biosocial Society and the Royal Anthropological Institute. We would also like to thank Lucy Radford who helped organise the workshop and the early stages of this volume, and all those at Berghahn Publishers who assisted in the development and production of this volume.

Catherine Hill
Amanda Webber
Nancy Priston

Introduction
Complex Problems
Using a Biosocial Approach to Understanding Human-Wildlife Interactions

Catherine M. Hill

The term 'human-wildlife conflict' is commonly used in the conservation literature to denote negative interactions between people and wildlife, i.e. where wildlife damage property including crops, or threaten the safety of livestock or even people. For many researchers interested in the conservation implications of these negative human-wildlife interactions, the entry point is a concern for wildlife. Consequently, the focus is often on what the animals do, and what people complain about. This perspective has, until very recently, dominated research and the design of conflict mitigation strategies. However, it is increasingly apparent that human-wildlife conflict is normally better understood as conflicts between different human groups, sometimes over how wildlife should be managed, but expressed as a clash between human and wildlife needs and activities (Madden 2004; Marshall, White and Anke 2007; Dickman 2010; Hill and Webber 2010; Redpath et al. 2013).

Increasingly researchers are labelling these human-wildlife conflicts as 'wicked problems' (e.g. Bal et al. 2011; Marchini 2014; chapters 1 and 8, this volume). A 'wicked problem' is one that is challenging or seemingly impossible to solve because of incomplete, contradictory and changing requirements that are often difficult to define. Such problems are characterized as multifaceted, involving multiple stakeholders who hold conflicting perspectives and values. Accordingly, these problems are hard to describe, tend to recur and may change in response to any attempt to solve them (Rittel and Webber 1984). Before such problems can be addressed there must first be some degree of consensus between interested parties as to what the problem is, and achieving this agreement can be a 'wicked problem' in itself (Rittel and Webber 1984). But are human-wildlife conflicts truly 'wicked' rather than simply complex?

A key feature of many human-wildlife conflicts is the involvement of multiple stakeholders whose priorities, perspectives and agendas are often incongruous, as illustrated clearly within this volume. These are common features of 'wicked problems'. However, mitigation attempts to date mostly focus on developing technical solutions to reduce the negative impacts of wildlife behaviour on human property or safety, without recognizing or addressing underlying social conflicts. As a consequence, they are rarely fully successful in addressing these 'conflicts' (see, for example, Webber, Hill and Reynolds 2007). Therefore, perhaps human-wildlife conflicts are perceived as 'wicked' problems because it is only very recently that their complex, biosocial nature has been recognized, and hence attempts to mitigate them have fallen short through a lack of understanding of this complexity rather than the majority of them necessarily being unsolvable. As pointed out by Peter Balint and colleagues, 'not all problems with multiple stakeholders and uncertain outcomes are wicked' (Balint et al. 2011: 30). Therefore, we should be cautious of labelling conflicts around wildlife as 'wicked', because it might encourage the view that these conflicts are unresolvable and thus it is impractical even to try addressing them, further jeopardizing human-wildlife coexistence.

The chapters in this volume span a variety of species, geographical locations and cultural contexts. The details of the case studies may be very specific, but between them they address several themes: that human-wildlife conflict is about power differentials between the different human protagonists and not necessarily about the wildlife per se, and that animals are important as symbols. The first of these themes is beginning to attract attention within mainstream conservation science; the second is largely ignored within this body of literature (Hill 2015), yet is highly apposite when exploring the challenges of facilitating human-wildlife coexistence.

A Biosocial Understanding

Relying solely on observations or reports of human behaviour towards or around animals may tell us relatively little about what animals really mean to people, why people interact with animals or even the value different human groups assign to animals, including wildlife. Instead we need to engage more fully with the ways people think and articulate about animals, animal behaviour or apparent competition between people and animal interests, because animals are 'good to think' with (Levi-Strauss 1963: 101) as well as good to eat, observe, admire and share space with. To develop a truly biosocial approach to exploring and understanding questions about human-wildlife interactions and human-wildlife coexistence requires careful integration of methodological and theoretical perspectives from both the natural and social sciences, and especially better understanding and acceptance of qualitative methods (Drury et al. 2011) and exploratory research models.

The challenge remains though as how best to promote and foster understandings and synergies between the different disciplinary perspectives, and particularly between quantitative and qualitative paradigms. Helen Newing (2013) suggests that the way forward to facilitate and foster research into biocultural diversity and conservation science is the development of scholars who have a basic understanding of other disciplinary ideas and approaches as well as the one they work in. Developing such capacity for truly transdisciplinary skills takes time. However, there is good evidence of increasing awareness and acceptance among natural scientists, social scientists and wildlife managers that to understand the complex, multifaceted nature of human-wildlife relationships and human-human relationships more fully, we need better integration between natural and social science perspectives (Edwards and Gibeau 2013; Inskip et al. 2014; White et al. 2009). Nevertheless, there are still only a few examples where interdisciplinary conflict mitigation is attempted and critically assessed (Dickman 2010). In this volume the various contributors explore these issues, providing unequivocal evidence of the complex nature of human-wildlife relationships, and the value of adopting a biosocial approach to understanding and managing human-wildlife interactions.

'Human Wildlife Conflicts': Do Labels Matter?

There is now a movement in the literature advocating that the term 'human-wildlife conflict' be dropped from common usage (Peterson et al. 2010; Redpath et al. 2013; Madden and McQuinn 2014; Hill 2015). Nils Peterson et al. argue that it is inaccurate as a label characterizing crop damage or livestock predation by wildlife because it implies 'conscious antagonism between wildlife and humans' (2010: 75). It also exaggerates the cognitive capacities of the animals, and masks the multifaceted, changeable nature of these conflicts that occur because of diverse values, priorities and power relations between the human stakeholder groups concerned. Framing these scenarios as human-wildlife conflict draws the focus away from the real identity of the protagonists, i.e. the various human stakeholders, thereby hindering the development of effective mitigation or resolution strategies. Additionally, it influences understandings and methodological approaches adopted in researching these scenarios by 'diverting attention from addressing conflicts within human political systems' (chapter 1, this volume). Accordingly, it focuses attention on changing the nature of the interaction between people and 'problem' wildlife, or changing local people's perceptions of, attitudes towards and willingness to share space and resources with wildlife. Where animal damage is labelled human-wildlife conflict it makes sense for people to direct their antagonism towards the animals involved, as 'perpetrators' of the 'conflict'. In some cases this can prompt retaliatory killings (Dickman 2010). Consequently, where people and wildlife are in competition over resources, the language used to describe these interactions, i.e., 'human-wildlife conflict' and the depiction of the animals concerned as 'pests', may exacerbate the problem and further endanger the long-term coexistence of people and wildlife.

Finding an alternative term to 'human-wildlife conflict' is proving problematic. Suggested alternative labels include human-wildlife coexistence (Madden 2004), human-human conflicts (Marshall, White and Anke 2007), human-wildlife competition (Matthiopoulos et al. 2008), conservation conflicts (Redpath et al. 2013) and human-wildlife interactions (Peterson et al. 2010). A preliminary survey of citations in Web of Science suggests that many authors publishing papers in this field of study are proving slow to adjust their terminology away from 'human-wildlife conflict' (Humle and Hill 2016). However, more careful analysis of the literature suggests the situation is not straightforward. Of a sample of 1,372 articles, accessed systematically through Science Direct, CAB Abstracts and PubMed, 60 per cent of articles published before 2001 adopted alternative terms to 'human-wildlife conflict' (Webber et al. forthcoming)[1]. By contrast, almost 70 per cent of articles published after 2001 (i.e. 2001–2015) used human-wildlife conflict language. This more recent uptake of such terminology appears linked to certain key publications, including Terry Messmer's 'Emergence of Human-Wildlife Conflict Management: Turning Challenges into Opportunities' (2000) and the International Union for the Conservation of Nature's (IUCN) 'World Parks Congress Recommendation: Preventing and Mitigating Human-Wildlife Conflicts (2003), that used human-wildlife conflict' expressions. Thus, 'human-wildlife conflict' has rapidly become a widely accepted term to denote an assortment of methodological approaches and conservation issues involving apparent competition between people and wildlife (Webber et al. forthcoming). Indeed, the variable terminology used by the authors within this volume is testament to the widespread use of the term, even where authors are specifically analysing 'human-wildlife conflicts' as fundamentally being conflicts between different human groups, rather than as direct conflict between human and animal protagonists.

What can a Biosocial Perspective Contribute?

In this book's first chapter, Phyllis Lee explores how farmers' views of wildlife species that damage crops compare with their perceptions of other animal species, particularly where animals are recognized as having economic, social or aesthetic value. She points out that while human experiences of wildlife behaviour may cause genuine conflicts of interest (e.g. crop losses through foraging or trampling; livestock losses to predators), it is important to be aware that different interest groups may represent 'conflicts' in specific ways to promote their own agendas, or even misrepresent others. This analysis of human-wildlife conflict as a phenomenon subject to political manoeuvrings on the part of different interest groups, creating competing and conflicting representations, priorities and expected and/or hoped for outcomes, resonates closely with later chapters, and particularly those by Ketil Skogen (chapter 3, on wolves in Norway), Angela Cassidy (chapter 4, badgers in the United Kingdom) and Francine Madden and Brian McQuinn (chapter 8, on conflict transformation). The basic tenet of Lee's chapter is that 'conflict' is not just about people-wildlife interactions but also interactions between people; therefore conflict scenarios must be understood within the relevant social contexts, including variable understandings, local value systems and consequent agendas.

In chapter 2, John Knight examines how Japanese farmers respond to the risk of crop damage by Japanese macaques (*Macaca fuscata*). Crop damage by this species is widespread and common in rural Japan, and its increased intensity and distribution attributed to the depopulation of rural areas (Sprague 2002). Indeed, rather than humans 'trespassing' on monkey spaces, it could be argued that monkeys are encroaching on human spaces because villages become less threatening environments as the number of human inhabitants declines. Knight argues that previously in Japan, crop damage was experienced as a 'human-monkey conflict', but as a consequence of human interference in 'the monkey problem' it is now viewed as a conflict between humans because it is through people's actions that monkeys have changed their behaviour and moved into human spaces. This is a different interpretation to that in other chapters in this volume (see chapters 1, 3, 4 and 8), because here it is the human actors, rather than the researcher, who acknowledge the 'conflict' as one between people rather than between people and macaques.

The Symbolic Nature of Animals

Interactions between humans and wildlife are neither simply a matter of direct physical encounters, nor of the exploitative uses that humans make of wildlife. Humans invest symbolic meaning in animals, and this is central to understanding the human-wildlife relationship. Individuals and groups may, for example, use animals as analogies in the theories and models they develop to explain human behaviour. Nevertheless, such analogies work both ways, influencing how the animals are themselves perceived, and how humans interact with them. In chapter 3, Ketil Skogen describes wolf management systems implemented in Norway, and how these systems neither satisfy the various human interest groups involved, nor provide adequate provision for wolves. He argues that conflicts over wolves go far beyond the economic impact or disputes over management options, and are about the threat of social change as perceived by rural communities. Skogen advises that for rural populations wolves are a symbol of rural decline because wolf presence is now associated with rural depopulation, deterioration in service provision in rural areas and abandonment of fields, which revert to 'nature'. Furthermore, for those involved in rural industries (agriculture, forestry), protection of species through conservation is symbolically associated with loss of control over how natural resources are managed by local people. Consequently, the success of wolves is symbolic of the decline of rural traditions and rural lives, with wolves valued by influential sectors of society above rural traditions, rural lives and therefore rural people.

By contrast, for the increasing numbers of middle class, well-educated people moving into rural areas the wolf is symbolic of 'an authentic, wild nature that preceded the human-dominated (and now partly damaged) landscape' (chapter 3). From Skogen's analysis it is clear that these 'conflicts' around wolves are a result of clashes between the entrenched views about the rural order held by different members of Norwegian society. Consequently, to defuse conflicts (and avoid precipitating new ones) managers need to be aware of the symbolic meanings assigned to wolves within Norwegian society, and understand that conflicts about wolves are a consequence of competing social constructions and value systems. What is really at stake here is not the actions of the wolf. Rather the underlying concern is whether rural Norway should be a wilderness that accommodates large carnivores, or a managed production landscape for timber and grazing, with little or no space for top predator species. This example illustrates clearly why conservationists and wildlife managers should be aware of the symbolic nature of animals because it affects how people might interpret conservationists' activities, value systems and priorities.

The symbolic investment of meaning in animals is also apparent in the way that people categorize different species, particularly wildlife species. The categories 'pest' and 'problem' animals are especially problematic, as highlighted by Lee (chapter 1) and Sushrut Jadhav and Maan Barua (2012) who suggest that an animal labelled as a 'crop raider' becomes a legitimate enemy. Once a species or animal is 'demonized' in this way it becomes much easier to justify exploiting or extirpating them because they are 'pests', particularly where pests are associated with adverse events, outcomes and consequences.

Competing Constructions as a Source of 'Conflict':
The Case of the 'Protected' Pest
As outlined above, animals take on many meanings and values for people. Consequently, it is useful to explore more fully the competing constructions of animals both within and between different interest groups. Angela Cassidy develops this theme in chapter 4, where she examines why culling badgers creates such intense reactions in the United Kingdom. Cassidy explores the dualistic framing of badgers in recent debates about bovine TB in the media, with 'good' badger as a sagacious, respected, woodland dweller, and 'bad' badger as predatory, destructive and a source of pestilence. She demonstrates how disputes over badger culling are 'intertwined with tensions between traditional British rural centres of power and modern urban elites', reminiscent of the foxhunting debate in the United Kingdom (Marvin 2000) and societal conflicts expressed through concerns about wolves in Norway (chapter 3). Cassidy reveals that these competing representations, 'good badger' and 'bad badger', are broadly aligned with environmental and agricultural framings of the bovine TB problem and therefore, anti- and pro-cull sympathies. Thus for some people badgers are pests to be eradicated; for others they are an iconic species of the UK countryside to be valued and protected.[2] In other words, the same animal can mean different things to different groups of people, illustrating the constructed nature of the category 'pest'. In this instance, human-wildlife conflict is not about the animal per se, or even its actions. Rather, it is a reflection of the socially constructed values or meanings a particular animal or species has for different interest groups. Similar findings are evident for a range of 'pest' species, including otters (Goedeke 2005), dingoes (Hytten 2009), possums (Wilks, Russell and Eymann 2008) and chimpanzees (Sousa 2014). So here we have the conundrum of 'protected' pests, animals that are afforded a degree of legal protection yet damage or threaten the safety of people, livestock, pets, crops and property (Knight 2000, 2008).

Animal 'Personhood'
Human-animal relationships are not necessarily simple or clear-cut. The way people perceive animals and understand their relationship with humans shapes their inter-

pretation and expectations of animals and how they behave. In chapter 5, Marc Brightman explores the implications of Amazonian ideas about personhood for human-animal interactions among the Trio of Southern Suriname. Traditionally, the Trio believe people, as babies, have to be moulded into becoming human, and this process continues to be reinforced throughout an individual's life through activities such as 'eating together'. The Trio extend this idea to animals, whereby animals that eat together regard themselves as human, and view humans as animals. Accepting that animals come into people's gardens (swiddens) to eat is, in essence, recognition of these animals' sentience and personhood, and therefore is a symbol of the shared humanity between the Trio and their wildlife neighbours. Brightman argues this idea of animals having personhood is central to understanding Trio responses to wildlife feeding on their crops. Outsiders 'looking in' might well label this 'human-wildlife conflict' but for the Trio there is nothing conflictual about it, providing they are able to 'eat well' with their kin.

Conflict Narratives as 'Weapons of the Weak'

James Scott, in his seminal work *Weapons of the Weak: Everyday Forms of Peasant Resistance*, proposed that everyday forms of 'resistance' can be viewed as political action (1985). The kinds of behaviours he was referring to as indicative of resistance include 'foot-dragging, dissimulations, false compliance, feigned ignorance, desertion, pilfering, smuggling, poaching, arson, slander, sabotage, surreptitious assault and murder, anonymous threats' (1989: 5). Such actions allow individuals or groups to express discontent or disagreement without drawing attention to themselves, thereby avoiding the risk of incurring negative consequences associated with overt dissent (Scott 1985). This acts as a mechanism by which low-ranking individuals may obstruct mechanisms that favour more influential members of the society (Scott 1989), and/or may 'act as a safety valve for social discontent' (Adas 1986: 82, cited in Korovkin 1999). As Lee suggests in chapter 1, farmer and livestock herder discourses of conflict could be better understood as a 'weapon of the weak'. People could be venting their fury on the animals for a variety of reasons, including because they feel unable to express or direct their anger and frustration towards the underlying causes of social 'conflict', i.e. other people such as wildlife authorities, researchers and even conservation groups. For example, small-scale farmers in Uganda express a sense of being powerless in situations where their interests and rights clash with those of officialdom and/or outsiders. So, these farmers may use a human-wildlife conflict framing as a vehicle for expressing anger, frustration and a sense of dispossession of autonomy without entering into direct conflict with authority figures that might prove damaging or threatening to them (Hill 2004). Indeed, human-wildlife conflict narratives could be understood as resistance to local conservation agendas, particularly if they become more prevalent with the arrival of outsiders and/or figures of authority. Furthermore, disenfranchised farmers, herders or even landowners may use human-wildlife conflict narratives as a coping mechanism – partly as a way of dealing with the nuisance value of wildlife but also as a way to resist the imposition of conservation ideas, projects, personnel or even the barrage of conservation narratives they are subject to that are often counter to their own interests, priorities and sense of equity.

Attitudes and Perceptions Are Not Necessarily Fixed

Attitudes are an important component of people's willingness or capacity to 'tolerate' sharing landscapes with wildlife, particularly predators (Treves and Bruskotter 2014), hence the idea that we need to identify, understand and change attitudes. However, research suggests that people's attitudes are shaped by underlying values and therefore tend to change slowly (Manfredo 2009; Heberlein 2012). Consequently, understanding more about the complex nature of attitudes and how, when and why they change may improve our understanding and capacity to manage 'conflict' situations more effectively.

The concept of Wildlife Value Orientations (WVO) draws on the cognitive hierarchy framework from social psychology, and provides a useful tool, and theoretical structure, for understanding different viewpoints about wildlife (Fulton, Manfredo and Lipscomb 1996). In chapter 6, Alia Dietsch, Michael Manfredo and Tara Teel affirm that people's reactions to human-wildlife conflict are chiefly determined by their underlying WVOs. Consequently, WVOs can be used to explore and predict likely public acceptance of 'conflict' mitigation strategies pre-implementation. Using the recovery of the wolf population in Washington state, the authors demonstrate there is a high level of support for wolf recovery in more urban areas where the likelihood of encountering wolves is very low and mutualism WVOs are prevalent. Mutualists are defined as those having an egalitarian ideology that extends ideas of social inclusion to animals, emphasizing animal equality and welfare (Wildavsky 1991, cited in Manfredo 2009). Mutualists view wildlife as having rights and as being something to be cared for. By contrast, people living in more rural areas, where the likelihood of wolf encounters is higher, express much lower support for wolf recovery. Here there are more people identified as Utilitarians. These people are characterized by being more accepting of lethal means of control of animal populations.

However, Dietsch, Manfredo and Teel demonstrate that WVOs are not fixed, and can change over time. The authors link societal changes in the United States with changing views towards wildlife and management options. For example, hunting has a long history of use as a wildlife-management strategy in the United States but is increasingly becoming unacceptable to particular groups of people (chapter 6). This could, for example, have significant implications for future management of deer populations if lethal options become much less acceptable to the wider public, causing increased numbers of deer-vehicle collisions and increased damage to agricultural crops. This switch from a domination to a mutualism orientation within the United States occurs in association with changing socioeconomic status and associated cultural change. While there is no guarantee that different societies will show similar responses of societal WVOs to increasing wealth, the authors identify this as an area worth exploring further, particularly in the context of countries like China that have a strongly Utilitarian view of wildlife. Perhaps then, as societies become increasingly wealthy and exposed to external views, societal WVOs will also change, with implications for wildlife conservation more globally.

Narratives of change feature strongly in chapter 7. Here Lisa Naughton-Treves and co-authors examine farmers' perceptions of crop loss to wildlife among people living around the edge of the national park, comparing the results of two data sets collected at two points in time, seventeen years apart. Animals are reportedly now travelling a little farther out from the park edge than they did seventeen years previously; some species are foraging in areas they did not visit formerly; and latterly wild pigs, previously labelled as highly troublesome, rarely figure in current 'conflict' narratives. Legislative changes have legalized hunting of wild pig, perhaps enhancing their status locally whereby they could be valued as a resource rather than regarded just as a 'pest' species. The same cannot be said of baboons however, who even though they can now legally be hunted, continue to be regarded with fear and loathing. Additionally, the authors report that irrespective of whether family farm locations have changed between the two study periods, family members were more likely to demonstrate consistency in the views they express than were unrelated respondents, irrespective of location, perhaps reflecting more closely shared WVOs among relatives.

The Way Forward: Conflict Mitigation or Conflict Transformation?

As outlined earlier, participatory processes are a necessity when delineating and addressing complex problems including those labelled 'human-wildlife conflicts'. Francine Madden and Brian McQuinn describe just such an approach to conflict mitigation in chapter 8. They explain how until very recently the focus in conservation efforts to reduce human-wildlife conflicts has been directed towards changing physical interactions between the animals and the people, rather than looking beyond the immediate, proximal evidence of conflict (e.g. complaints about crop damage or livestock losses) and identifying, acknowledging and addressing underlying social conflicts between different human groups. Madden and McQuinn discuss the limitations of prioritizing interventions focused solely on technical solutions promoting change to human or animal behaviour. They suggest an alternative model for identifying the different types or levels of conflict, and describe an innovative approach, Conservation Conflict Transformation (CCT), which is currently being adopted across a range of different wildlife-related conflict scenarios. This approach fosters a more nuanced analysis of conflict, and the recognition that different types of social conflicts may affect mitigation processes differently. Consequently, the CCT approach facilitates better understandings of the underlying causes of conflict and their visible manifestations, and helps identify when conflict narratives are likely to be a proxy for the expression of underlying or deep-rooted social conflicts (Madden and McQuinn 2014).

A common assumption within the literature is that by reducing the negative impacts and nuisance aspects created by wildlife in shared landscapes, one fosters and encourages people's tolerance of wildlife – i.e. one promotes coexistence.

However, a failure to recognize or acknowledge existing intra-human social conflicts may mean that tools or procedures to reduce nuisance aspects of sharing landscapes with wildlife, however effective and easily applied, do not necessarily reduce local conflict rhetoric or conflict experience. As Madden and Quinn argue, existing intra-human conflicts need to be resolved before shifting to focus on technical solutions. Such an approach is more likely to get better community buy-in to implementing and maintaining technical solutions, therefore implementing CCT where appropriate, as a precursor to applying technical solutions, is likely to enhance their effectiveness and their uptake locally.

However, technical approaches that reduce the practical challenges of sharing landscapes with wildlife do have their place. Careful analysis of livestock husbandry practice, crop choice and planting patterns, proximity to wildlife refuges and wildlife behaviour will highlight points within livestock and farming systems that render domestic animals or crops vulnerable to predation by certain species. With those things in mind, it is often possible to suggest alternate husbandry strategies or farming practices, i.e. technical interventions that effectively reduce losses to wildlife (Jackson and Wangchuck 2004; Graham and Ochieng 2008; Davies et al. 2011; Hill and Wallace 2012; Potgeiter et al. 2013). However, such approaches should take into account additional factors relating to target user groups, including their priorities and concerns, competing labour requirements, cultural/social factors and user-group expectations of outcomes (Hill 2004; Webber, Hill and Reynolds 2007). Accordingly, technical solutions to reduce crop or livestock losses for example, should take account of what farmers expect of any intervention, making it clear from the outset who will be responsible for implementation and care of deterrents afterwards. Otherwise, irrespective of the effectiveness of any tool or strategy, where end-user groups' priorities and expectations are neither understood nor accommodated, any intervention might fail to reduce crop or livestock losses for example, through lack of uptake and engagement by farmers (Hill 2004; Webber, Hill and Reynolds 2007; Hill and Wallace 2012). Consequently, deterrent value is also affected by user opinions and expectations as to how likely they are to work, as well as by opportunity costs involved in having to set up, use and support any technique (Osborn and Hill 2005; Graham and Ochieng 2008). The last two chapters of this volume discuss the challenges of engaging with end-user groups to ensure their adoption of technical solutions, and consider how to scale these responses up for effective management of existing issues and prevention of future problems at the landscape level.

In chapter 9 Graham Wallace and Catherine Hill describe the process of developing a series of crop-protection tools (e.g. fencing, early warning systems and chemical repellents) in partnership with small-scale farmers in Uganda. Additionally, they reflect on factors influencing the degree to which farmers take ownership of such interventions. All farmers involved in the study requested that installations remain on their fields at the end of the project. Most of these installations were in use a year later and, in addition, neighbouring farmers not in the original study were recorded having installed similar or modified versions of the trialled deterrents on their own farms (Hsiao et al. 2013). Farmer engagement was achieved by involving end-user groups at all stages of the project, from identifying key issues, design of tools, data collection and evaluation of their utility. Such involvement on the part of the farmers ensures that tools and techniques address their concerns and priorities and methods are locally acceptable and manageable, and it encourages farmers to assume ownership of the project and thus responsibility for its evaluation, extended use and further refinement and development within the wider community.

In the final chapter Amanda Webber and colleagues explore the uses of Geographic Information Systems (GIS) tools to help researchers, local people and wildlife agencies predict areas vulnerable to wildlife foraging activity, identify key locations in which to focus crop-protection activities and visualize the 'problem' through maps of risk hot spots. They point out that risk maps generated through GIS data are useful as visual prompts to elicit discussion of issues and likely solutions. In this way, GIS becomes more than just a mapping technology and actively contributes to encouraging stakeholder engagement, shifting the focus from 'problem' to 'solution'. Simultaneously, it can be used as a tool to encourage more open discussion and consideration of the different stakeholder viewpoints as part of a trust and respect-building process, both of which are central components to 'transforming conflict', as per Madden and McQuinn (chapter 8).

Conclusion

This edited collection reflects our insights as researchers who over the years have shifted our thinking to an approach more firmly embedded within the social sciences. This shift in focus facilitates a more detailed and nuanced understanding of the relevant issues, both at individual and societal levels. As demonstrated in this volume, the nature of these conflicts around wildlife is complex and stems from misunderstandings, lack of awareness, acknowledgement and respect for alternative viewpoints, value systems, priorities and needs. However, to fully understand and manage, mitigate or even transform these 'conflicts' requires close examination of different scenarios through a biosocial lens. Technical solutions can be applied in tandem with conflict transformation processes, to reduce absolute costs to people of sharing landscapes with wildlife, but even these warrant a more detailed biosocial approach that incorporates careful analysis of wildlife ecology and human behaviour to develop effective and humane methods of protecting human property from wildlife actions. As stated by Madden and McQuinn (chapter 8), 'The future of conservation for many wildlife species relies not just on innovative solutions, but also on an increased tolerance and social carrying capacity that cannot be achieved by laws, science, money, or fences alone'. With this book we hope to contribute to this by encouraging a radical change in understanding, expectations and approaches as adopted by researchers, wildlife managers, conservationists, policy makers and funders alike, all of whom influence understandings, priorities, interventions and outcomes.

Catherine Hill is Professor of Anthropology in the Department of Social Sciences at Oxford Brookes University. Her research focuses on people-wildlife relationships, conservation conflicts and implications for people's perceptions of wildlife, biodiversity conservation and local communities.

Notes

1. Human-wildlife conflict terms monitored were 'human-wildlife conflict/s', 'crop raid/er/s/ing' and 'human-animal conflict/s'. Alternative terms monitored were 'conservation conflict/s', 'human-human conflict/s', 'human-wildlife interaction/s', 'human-wildlife coexistence', 'human-wildlife relationship/s', 'human-wildlife competition' and 'human-wildlife impact'.
2. The European badger (*Meles meles*) is protected in England and Wales under the Protection of Badgers Act 1992.

References

Bal, P., et al. 2011. 'Elephants Also Like Coffee: Trends and Drivers of Human–Elephant Conflicts in Coffee Agroforestry Landscapes of Kodagu, Western Ghats, India', *Environmental Management*. doi:10.1007/s00267–011–9636–1.

Balint, P.J., et al. 2011. *Wicked Environmental Problems*. Washington, DC: Island Press.

Davies, T.E., et al. 2011. 'Effectiveness of Intervention Methods Against Crop Raiding Elephants', *Conservation Letters* 4(5): 346–354.

Dickman, A.J. 2010. 'Complexities of Conflict: The Importance of Considering Social Factors for Effectively Resolving Human-Wildlife Conflict', *Animal Conservation* 13(5): 458–466.

Drury, R., et al. 2011. 'Less is More: The Potential of Qualitative Approaches in Conservation Research', *Animal Conservation* 14: 18–24.

Edwards, F.N., and M.L. Gibeau. 2013. 'Engaging People in Meaningful Problem Solving', *Conservation Biology* 27: 239–241.

Fulton, D.C., M.J. Manfredo and J. Lipscomb. 1996: 'Wildlife Value Orientations: A Conceptual and Measurement Approach', *Human Dimensions of Wildlife* 1: 24–47.

Goedeke, T.L. 2005. 'Devils, Angels or Animals: The Social Construction of Otters in Conflict over Management', in A. Herda-Rapp and T.L. Goedeke (eds), *Mad About Wildlife: Looking at Social Conflict over Wildlife*. Leiden and Boston, MA: Brill, pp. 25–50.

Graham, M.D., and T. Ochieng. 2008. 'Uptake and Performance of Farm-Based Measures for Reducing Crop Raiding by Elephants *Loxodonta africana* among Smallholder Farms in Laikipia District, Kenya', *Oryx* 42(1): 76–82.

Heberlein, T. 2012. 'Navigating Environmental Attitudes', *Conservation Biology* 26(4): 583–585.

Hill, C.M. 2004. 'Farmers' Perspectives of Conflict at the Wildlife-Agriculture Boundary; An African Case Study', *Journal of Human Dimensions of Wildlife* 9(4): 279–286.

———. 2015. 'Perspectives of "Conflict" at the Wildlife-Agriculture Boundary: 10 Years On', *Human Dimensions of Wildlife* 20(4). doi:10.1080/10871209.2015.1004143.

Hill, C.M., and G.E. Wallace. 2012. 'Crop Protection and Conflict Mitigation: Reducing the Costs of Living alongside Non-Human Primates', *Biodiversity and Conservation* 21: 2569–2587.

Hill, C.M., and A.D. Webber. 2010. 'Perceptions of Nonhuman Primates in Human-Wildlife Conflict Scenarios', *American Journal of Primatology* 72: 912–924.

Hsaio, S., et al. 2013. 'Crop Raiding Deterrents around Budongo Forest Reserve: An Evaluation through Farmer Actions and Perceptions', *Oryx* 47: 569–577.

Humle, T., and C.M. Hill. 2016. 'People-Primate Interactions: Implications for Primate Conservation', in S. Wich and A. Marshall (eds), *An Introduction to Primate Conservation*. Oxford: Oxford University Press, Pp 219–240.

Hytten, K.F. 2009. 'Dingo Dualisms: Exploring the Ambiguous Identity of Australian Dingoes', *Australian Zoologist* 35(1): 18–27.

Inskip, C., et al. 2014. 'Understanding Carnivore Killing Behaviour: Exploring the Motivations for Tiger Killing in the Sundarbans, Bangladesh', *Biological Conservation* 180: 42–50.

IUCN. 2003. 'World Parks Congress Recommendation: Preventing and Mitigating Human–Wildlife Conflicts.' Reprinted in *Human Dimensions of Wildlife* 9: 259–260 (2004).

Jackson, R.M., and R. Wangchuck. 2004. 'A Community-Based Approach to Mitigating Livestock Predation by Snow Leopards', *Human Dimensions of Wildlife* 9(4): 307–315.

Jadhav, S., and M. Barua. 2012. 'The Elephant Vanishes: Impact of Human-Elephant Conflict on Human Well-Being', *Health and Place* 18: 1356–1365.

Knight, C. 2008. 'The Bear as "Endangered Pest": Symbolism and Paradox in Newspaper Coverage of the "Bear Problem"', *Japan Forum* 20(2): 171–192.

Knight, J. 2000. 'Introduction', in J. Knight (ed.), *Natural Enemies: People-Wildlife Conflicts in Anthropological Perspective*. London: Routledge, pp. 1–35.

Korovkin, T. 1999. *Weak Weapons, Strong Weapons? Hidden Resistance and Political Protest in Highland Ecuador*. CERLAC Working Paper Series, November 1999.

Levi-Strauss, C. 1963. *Totemism*, trans. R. Needham. Boston: Beacon.

Madden, F. 2004. 'Creating Coexistence between Humans and Wildlife: Global Perspectives on Local Efforts to Address Human-Wildlife Conflict', *Human Dimensions of Wildlife* 9: 247–257.

Madden, F., and F. McQuinn. 2014. 'Conservation's Blind Spot: The Case for Conflict Transformation in Wildlife Conservation', *Biological Conservation* 178: 97–106.

Manfredo, M.J. 2009. *Who Cares About Wildlife? Social Science Concepts for Studying Human-Wildlife Relationships and Conservation Issues*. New York: Springer Press.

Marchini, S. 2014. 'Who's in Conflict With Whom? Human Dimensions of the Conflicts Involving Wildlife', in L.M.Verdade, M.C. Lyra-Jorge and C.I. Piña (eds), *Applied Ecology and Human Dimensions in Biological Conservation*. Heidelberg, New York, Dordrecht and London: Springer, pp. 189–209.

Marshall, K., R. White and F. Anke. 2007. 'Conflicts between Humans over Wildlife Management: On the Diversity of Stakeholder Attitudes and Implications for Conflict Management', *Biodiversity Conservation* 16: 3129–3146.

Marvin, G. 2000. 'The Problem of Foxes: Legitimate and Illegitimate Killing in the English Countryside', in J. Knight (ed.), *Natural Enemies: People-Wildlife Conflicts in Anthropological Perspective*. London: Routledge, pp. 189–211.

Matthiopoulos, J., et al. 2008. 'Getting below the Surface of Marine Mammal-Fisheries Competition', *Mammal Review* 38: 167–188.

Messmer, T.A. 2000. 'Emergence of Human–Wildlife Conflict Management: Turning Challenges into Opportunities', *International Biodeterioration* 45: 97–100.

Newing, H. 2013. 'Bridging the Gap: Interdisciplinarity, Biocultural Diversity and Conservation', in S. Pilgrim and J. Pretty (eds), *Nature and Culture*. London: Earthscan, pp. 23–40.

Osborn, F.V., and C.M. Hill. 2005. 'Techniques to Reduce Crop Loss to Elephants and Primates in Africa: The Human and Technical Dimension', in R. Woodroffe, S. Thirgood and A. Rabinowitz (eds), *People and Wildlife: Conflict or Co-existence?* Cambridge: Cambridge University Press, pp. 72–85.

Peterson, M.N., et al. 2010. 'Rearticulating the Myth of Human-Wildlife Conflict', *Conservation Letters* 3: 74–82.

Potgeiter, G.C., et al. 2013. 'Why Namibian Farmers are Satisfied with the Performance of Their Livestock Guarding Dogs', *Human Dimensions of Wildlife* 18(6): 403–415.

Redpath, S.M., et al. 2013. 'Understanding and Managing Conservation Conflicts', *Trends in Ecology & Evolution* 28(2): 100–109.

Rittel, H.W.J., and M.M. Webber. 1984. 'Planning Problems are Wicked Problems', in N. Cross (ed.), *Developments in Design Methodology*. Chichester: John Wiley & Sons, pp. 135–144.

Scott, J.C. 1985. *Weapons of the Weak: Everyday Forms of Peasant Resistance*. New Haven, CT, and London: Yale University Press.

———. 1989. 'Everyday Forms of Resistance', in F.D. Colburn (ed.), *Everyday Forms of Peasant Resistance*. Armonk, NY: M.E. Sharpe, pp. 3–33.

Sousa, J. 2014. 'Shape-Shifting Nature in a Contested Landscape of Southern Guinea-Bissau', PhD thesis, Oxford Brookes University.

Sprague, D.S. 2002. 'Monkeys in the Backyard: Encroaching Wildlife and Rural Communities in Japan', in A. Fuentes and L. Wolfe (eds), *Primates Face to Face: The Conservation Implications of Human-Nonhuman Primate Interconnections*. Cambridge: Cambridge University Press, pp. 254–272.

Treves, A., and J.T. Bruskotter. 2014. 'Tolerance for Predatory Wildlife', *Science* 344(6183): 476–477.

Webber, A.D., C.M. Hill and A. Acerbi. Forthcoming. 'Examining the Language of Conservation: The Diffusion of "Human-Wildlife Conflict" and Alternative Terms in the Scientific Literature.'

Webber, A.D., C.M. Hill and V. Reynolds. 2007. 'Assessing the Failure of a Community-Based Human-Wildlife Conflict Mitigation Project in Budongo Forest Reserve, Uganda', *Oryx* 41(2): 177–184.

White, R.M., et al., 2009. 'Developing an Integrated Conceptual Framework to Understand Biodiversity Conflicts', *Land Use Policy* 26: 242–253.

Wilks, S., T. Russell and J. Eymann. 2008. 'Valued Guest or Vilified Pest? How Attitudes towards Urban Brushtail Possums *Trichosurus vulpecula* Fit into General Perceptions of Animals', in D. Lunney, A.J. Munn and W. Meikle (eds), *Too Close for Comfort: Contentious Issues in Human-Wildlife Encounters*. Royal Zoological Society of New South Wales, pp. 285–295.

1
People, Perceptions and 'Pests'
Human-Wildlife Interactions and the Politics of Conflict

Phyllis C. Lee

It is a tautology to state that human livelihood interactions with our natural environments are almost universally destructive and detrimental to ecosystem persistence. However, if we as researchers aim to manage or mitigate these human destructive capacities effectively, we need to understand the quality and nature of local attitudes to ecosystems and in particular protected areas, unique habitats or species of conservation concern. Negative perceptions of externally legislated protection can arise from restrictions on the use of plants, animals, water or localities, or result from direct interactions with wildlife and wildlife's impact on local livelihoods, or both. It is also clear that gender, ethnicity and cultural expectations and values underlie attitudes and actions on the part of both local people and the socially, politically and financially empowered conservationists. It is often argued that providing evidence of the potential benefits of biodiversity conservation – via education and livelihood support – might enhance compliance with protection, but as yet, explorations of this construct provide conflicting answers (Murphree 1993; Ferraro and Kiss 2002; Flinton 2003; Naidoo and Ricketts 2006).

The goal of this chapter is to examine several case studies that have explored attitudes towards areas and animal species of conservation concern (both protected and unprotected). I also aim to assess some of the values that people report as applying to animals, which can translate into hostility or negative actions towards wildlife, especially when there is a cost to human livelihoods as a result of contact and interaction. In common parlance, contacts, interactions and hostile perceptions have long been labelled as 'conflict', and this labelling itself – the use of linguistic constructs to establish people's and practitioners' approaches to wildlife – is a significant issue in the discourse of conservation. However, the effects of the use of the 'conflict' paradigm are less often examined by the biodiversity conservationists who coined the term (see Peterson et al. 2010). Examining human-wildlife interaction both within and outwith these conflict paradigms might help predict contexts when coexistence, or at least tolerance, can be enabled between the human and non-human interactants (Fuentes and Wolfe 2002; Riley 2006).

Several factors can be considered in order to separate the phenomena underlying the attitudes of people interacting with wildlife. Many such explorations require some degree of cost-benefit analysis, but not necessarily a financial one. These cost-benefit analyses take the form of understanding the nature of the markets in which the local humans are operating: are the economies based on extrinsic values (commoditized: Swift, Izac and van Noordwijk 2004; Igoe and Brockington 2007; Vira and Adams 2009) or intrinsic (aesthetic, cultural: Kellert 1996; Kuriyan 2002; Riley 2006) ones? Then we can ask if there is a larger context to a local so-called conflict: is there a global exchange that is being maximized or which will externally impact on local exchanges? Globalized market opportunities can enhance perceived conflict. For example, problem animals or parts of these can be traded at huge value to the benefit of only a small number of individuals – e.g. the traders and importers of ivory, tiger bones or rhinoceros horns (Bennett 2011) – while the larger local community suffers the opportunity costs of lost biodiversity and the risks of being drawn into an illegal hunting trade. We also need to ask where power resides in communities – with equality across local actors, or stratified by status, wealth or gender? And then, what actions are available for conflict management; are these externally imposed and controlled or self-generated and self-policed (Haenn 1999; Zimmerman et al. 2001; Barros, Pereira and Vincente 2011)? The removal of the capacity for local decision-making again can exacerbate perceptions of conflict given a lack of local ability to control so-called[1] problem animals.

Finally, what other external factors lie at the root of conflict: publicity, power, political manipulations of people and their land or exploitation of local discontent to gain political power? In other words, while human interaction with animals may result in genuine livelihood conflict at the local level, the context of conflict can be manipulated for political benefits at a considerable distance from those actually experiencing the negative interactions. The construct of political and economic systems as the interconnected codification of social relations among human communities does, however, need to separate the levels at which these relations and social exchanges occur; individually, locally, regionally, nationally or globally. It is not always clear where conflict intervenes in shaping these larger relationships. Is conflict social in origin, reflecting unhappy social changes and personal power imbalances, or is it conflict between competing values and unshared goals within a society?

Politics and power establish the context of conflict as these aspects determine conservation policies and priorities at all levels, from the Convention of Parties meetings for CITES (Convention on International Trade in Endangered Species of Wild Fauna and Flora) and the very classification of species as threatened, to the gazetting and legislation associated with protected areas – whether in the United

States, the European Union or developing regions. Therefore, understanding how politics – from the level of gender, religion, local governance, regional and global – impact on people's perceptions is vital to understanding the potential outcomes of human-animal interactions. As noted above, local conflict can translate into regional or national political debates, and, as has been widely discussed, the so-called benefits of conservation (biodiversity maintenance, ecosystem services, wealth generation through ecotourism, protection of cultural heritage, etc.) also accrue at the regional or global rather than the immediate, individual or local level. Making arguments about such global conservation benefits to people directly experiencing impacts from wildlife may not change their negative perceptions of wildlife, or of the conservationists who protect wildlife to the apparent detriment of local residents. Conflict exists therefore at a number of levels, each of which needs to be considered both separately and within the expanded contexts (figure 1.1).

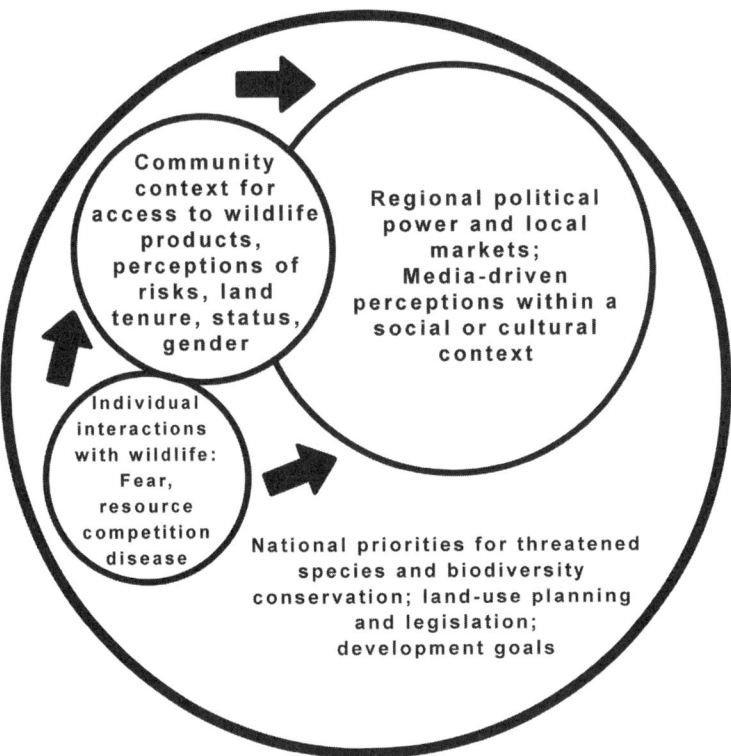

Figure 1.1. Circles of hierarchical and interacting contexts underlying and determining people's perception of 'conflict' with wildlife. Economic, social and political power are indicated by the size of the circle, while the flow of financial incentives and benefits (indicated by the arrows) tends to be outwards from the smaller circles to the national, leaving local people to share only the risks and costs. At each level, the media – local or global – can exploit issues for their own financial or political gain.

Although the discussion is framed in terms of costs and benefits (individual or global), perceptions and attitudes derive from more than simple trade-offs. Human psychology has been considered as rooted in the concept of self, selfishness, market-driven exchanges or inequality aversion (Soltis, Boyd and Richerson 1995; Bowles and Gintis 2004; Flack, de Waal and Krakauer 2005; Humphrey 2007), but we are equally likely from an early age to expect and act in a cooperative manner, tolerating inequality for some perceived generalized benefits (Tomasello et al. 2005). Conservation can work, but we must understand the drivers in each local context, and build upon this knowledge. Each site-specific 'ethno-ecology' will determine the local policies (Haenn 1999; Barros, Pereira and Vincente 2011), but we must understand the underlying ethno-ecology first.

Attitudes represent individually and then socially embedded and constructed beliefs about the environment and other species, what is known about environments and other species, and potentially reflect the interaction between knowledge and belief. Thus attitudes represent elements of identity, socialization to values, norms and beliefs, environmental knowledge, actions, intentions *and* conflict (Giddens, Duneier and Appelbaum 2003; Browne-Nuñez and Jonker 2008; Kellert 2008). I will try to tease apart some of these social and biological constructs affecting attitudes and examine conflict through the use of several specific case studies within a broader context of habitats and species of conservation concern. In these case studies, a variety of social science methods have been used – from quantitative assessments of values and willingness to make trade-offs (e.g. Gregory, Lichtenstein and Slovic 1993; Baron and Spranca 1997; Sagoff 1998), to qualitative explorations of associations between constructs or statements of values, to explorations of ethno-ecological knowledge (e.g. Becker and Ghimire 2003; Borgerhoff Mulder et al. 2009; Sitati and Ipara 2012).

Perceptions and Attitudes to Biodiversity Initiatives

Explorations of attitudes towards conservation or perceptions of the value of sustaining or protecting biodiversity in emerging economies with people experiencing constraints on livelihoods due to wildlife still come up with a roughly 50:50 (positive:negative) split in a population (Hill 1998; Mehta and Heinen 2001; Kuriyan 2002; Gillingham and Lee 2003; Bandara and Tisdall 2004; Gadd 2005; Lindsay, du Toit and Mills 2005; Baral and Heinen 2008; Karanth et al. 2008). These authors show that positive responses about biodiversity protection or attitudes reflecting environmental concerns are associated with the local provision of information and good mechanisms of governance. In addition, these relate to aesthetics, the real or perceived benefits of employment or revenue generation, and to the protection of crops and/or people. Negative perceptions reflect both economic or livelihood losses and conflicts of interest over power and values, and are often associated with heavy-handed implementation of compliance with externally mandated protected area or species legislation. Common complaints are a lack of local or individual participation in decision-making over resource management, and a general lack of support for, or willingness to participate with and trust, the external institutions that are responsible for the implementation of conservation strategies. In keeping with global analyses of attitudes, individual (or egocentric) values can trump biospheric (environmental) and altruistic values (e.g. Schultz et al. 2005), especially when behavioural changes or costly actions are required by those individuals.

Broad-scale understanding and synthesis of how perceptions shape actions and reactions to human-wildlife interactions across a variety of contexts are needed if we are to manage conflicts (e.g. Madden 2004; Fisher 2016) and ensure the persistence of threatened ecosystems or species. I focus here on conservation outcomes, identifying issues that generated conflict or alternatively activities that effected conflict resolution. I compare results from three specific studies, two in East Africa and the other in West Africa, on perceptions of protected areas by local inhabitants who also suffered crop-raiding by a variety of wildlife species from birds to buffalo and chimpanzees to elephants, and place these into a broader context. Our work in Tanzania and Guinea-Bissau address attitudes in the context of local politics and gender, while the Kenyan study focuses specifically on one charismatic megafauna, the elephant, which has become an icon of human-wildlife conflict across its range (Hoare 2000; Lee and Graham 2006). The persistence of elephants into the future looks increasingly imperilled when 'conflict' provides a rationale for rampant consumptive utilization and as the global market for illegal ivory expands exponentially.

Conservation Attitudes: Selous Game Reserve, Tanzania

Attitudes to species conservation, protected-area management and economic incentives were explored during the development of a community-based wildlife-management plan in the Mgeta River buffer zone, Selous Game Reserve, Tanzania (Gillingham 1998). The aim of the management plan was to develop strategies to minimize conflict with crop-raiding wildlife, and to provide economic incentives to the local residents as livelihood compensation for living alongside wildlife as well as for the recent restrictions on bushmeat hunting or the use of other wildlife and plant products. Sarah Gillingham and Lee (1999; 2003) found that people's perceptions of the protected area and the wildlife were gendered as well as a function of local social status and the local political context (see table 1.1). This local political context depended on who controlled the meat and hunting revenue distributions, which were seen as unequal. As noted above, humans have highly sensitive psychological mechanisms for detecting inequality, which is seldom taken into account when establishing revenue-sharing schemes where inequality can be easily reinforced. A further issue was that of restrictions on land ownership (possibly due to Tanzania's historical policies of 'villagization' or collective resettlements). General support for conservation was voiced, but this support was devolved to the national rather than the local level. In common with many politically marginalized peoples (Newton 2001), benefits were perceived to accrue at the national rather than individual or local levels even when projects were explicitly designed to provide benefits locally. Passive recipients in such schemes can become marginalized from the aims of the schemes. Finally, social capital (community identity) and collective actions were lacking with regard to biodiversity conservation even in small rural villages in a country with a history of social equality and communal land ownership. Co-management of wildlife clearly required accountable local institutions with a well-defined collective identity.

Factors underlying expressed attitudes – both positive and negative – can be compared across studies (table 1.1); influences on attitudes to protected areas among rural smallholders living in these areas were contrasted with attitudes among smallholders and pastoralists experiencing human-elephant conflict outside a protected area. As noted earlier, positive attitudes did emerge but protected areas were potentially viewed negatively as was one specific protected but crop-raiding species, the elephant. What is common to both contexts is the distinction already noted between individual costs incurred in a conservation context (egocentric values), and the recognition that benefits can accrue at higher social levels or across a larger landscape (biospheric values).

Table 1.1. Synopsis of sociocultural factors affecting attitudes from two studies in East Africa.

A. Perceptions of Protected Areas (Gillingham and Lee 2003)

Factor	Attitude	Predictive Factor
Knowledge of injury / death	Negative	Recent experience; connectedness of experience to respondent
Experience with the need to deter elephants from crops	Negative	Effort and expense involved
Area under crops	Negative	Smallholders
	Positive	Large landholdings with diverse activities
Likelihood of contact	Negative	Frequent persistent contacts
	Positive	Rare contacts

B. Perceptions of Pests (Elephants) (Graham 2007)

Factor	Attitude	Predictive Factor
Personal benefits (e.g. meat quota or hunting funds)	Positive	Higher status, greater wealth
	Negative	Gender (women)
State benefits (tourism, hunting fees, global conservation)	Negative	Gender (women)
	Positive	More education
Community engagement	Negative	Gender (women)
	Positive	Higher status, perceived personal benefits

Perceptions of Species and Conservation Attitudes: Guinea-Bissau

While there has been a long tradition of attitude studies in the developing economies of East/Southern Africa (see Browne-Nuñez and Jonker 2008), India (Karanth et al. 2008), Nepal (Baral and Heinen 2008) and South America (Borgerhoff Mulder et al. 2009), fewer studies of people's perceptions have focused on West Africa. Recent assessments of initiatives to protect unique Guinean biodiversity in Tombali and Lagoas de Cufada Natural Park, Guinea-Bissau (Costa et al. 2008; Costa 2011; Hockings and Sousa 2011) have illustrated the attitudinal consequences of the establishment of protected areas with villages embedded within the protected habitats and livelihoods that are completely dependent on exploiting the local ecological infrastructure. In these areas, bushmeat poaching is devastating mammalian biodiversity, slash and burn agriculture is causing serious deforestation, larger plantations (rice, cashews) expand with unaccountable biological management and in addition there is extensive crop-raiding, especially by endangered chimpanzees (*Pan troglodytes verus*) along with other non-human primates of conservation concern. The chimpanzees were initially proposed by conservation NGOs in the area as a potential flagship species to promote positive perceptions of biodiversity conservation. Susanna Costa (2011) applied the construct of a hierarchical system of beliefs (Arnold Arluke and Clinton Sanders's [1996] 'socio-zoological scale') about the values that different wild and domestic species had for humans (figure 1.2) as a mechanism to assess

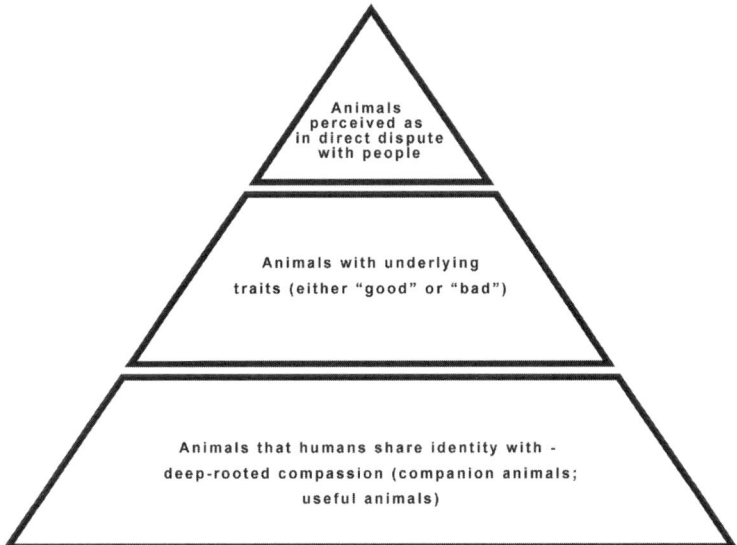

Figure 1.2. The concept of the socio-zoological scale; adapted and modified from Arluke and Sanders (1996) and Costa (2011). The simple hierarchical structure represents an attempt to access beliefs about different species based on qualities and values attributed to these species by people, and assumes that there is a relatively deterministic categorization (good/bad) in operation to organize these beliefs.

perceptions, as well as working with ecological knowledge on the part of hunters and perceptions of conflict from focus groups of women farmers. Similar to other studies exploring the values attached to wild species (Kaltenborn et al. 2006) a range of preferences and dislikes was found, independent of the risks posed by the species. When combined with risks, these values influenced the acceptability of management and protection options for each of the different species. It became clear that chimpanzees were *not* a good model species for a conservation flagship in this region of Guinea-Bissau. They were perceived of as similar to humans (which was not entirely positive, being cunning) and by comparison to other non-human primates and edible ungulates, people held few positive values about chimpanzees, as they were considered inedible and aesthetically ugly. They were also feared for their aggressive crop-raiding of fruit trees essential to the local market economy.

Costa's (2011) study also examined livelihood constraints based on risk mapping and she identified factors and events perceived of as beyond people's control (famines and crop-failures). There was no stated recognition during focus group risk mapping that people's activities – deforestation and hunting – were affecting land productivity and biodiversity, which were implicated in crop failures and disease (table 1.2A). Market economics were increasing in importance in this region due to changing requirements for cash (e.g. the need to sustain communication networks that underlie family and financial relations, such as through credit for mobile phones), driving a shift from subsistence bushmeat hunting to an unsustainable regional trade as well as exacerbating intolerance of crop losses due to wildlife. International issues of economics were also of significance in that rural Guinea-Bissau has become a hub for global cocaine distribution (UNODC 2007), which has the potential to skew local economies out of all recognition.

What conclusions about attitudes to conflict can be reached as a result of these studies in Guinea-Bissau? Conflict perception derives from direct interaction with crop-raiders (individual level) and also from constraints on livelihoods (hunting, fishing and resource extraction; community economic level). Local solutions to local livelihood constraints are required but there are questions as to how these can emerge and who will pay for them and monitor their success. Reinforcement of local education (village schools) provided by parks as a social benefit, including education for adult men and women in a society with very low levels of literacy, could lead to more positive choices about the use of resources. Ecotourism, while stated to be of some potential interest to some men in the surveys, might not be either possible or sustainable, given the lack of any transport or hospitality infrastructure in these regions. Political and civil instability or disease outbreaks (e.g. Ebola or cholera) at the national level are also barriers to developing wildlife tourism opportunities. Conflict zones, local coups or even publicized events of terrorism, warfare or disease in adjacent regions tend to discourage all but the most intrepid tourists.

Table 1.2. Comparison of livelihood risk analysis for people in conflict contexts; risk mapping based on participatory interviews (e.g. Smith, Barrett and Box 2000) or weighted rank index of household reports. Risk index values range between 0 and 1, with 1 being indicative of a highly ranked risk. *Source*: adapted from Costa 2011 (A) and Graham 2007 (B).

A. **Risk Perceptions Index, Tombali, Guinea-Bissau, from Discourse Analysis (three focus groups with 47 Women, 47 conversations with men: Costa 2011)**

Factor	Severity index	Incidence index	Risk index
Famine	1.14	0.8	0.7
Health	1.46	0.28	0.2
Money	1.65	0.28	0.13
Water	1.42	0.13	0.09
Other*	1.6	0.11	0.07
Housing	1.38	0.09	0.06

* Transportation, schools, roads etc.

B. **Weighted Risk Index, Laikipia, Kenya (open-ended responses to fixed question; n = 352 households, Graham 2007)**

Factor	Weighted Rank Index	WRI Order	% of households reporting factors	% of households ranking factors 1st
Drought	0.63	1	87.9	41.7
Disease	0.59	2	82.3	37.8
Wildlife	0.27	3	61.3	9.5
Cattle Rustling	0.14	4	32.7	5
Other*	0.05	5	15.1	0.8

* Fire, poaching, grazing by outsiders

Perceptions and Attitudes to Species: The Elephant Example

Turning to the issue of direct and dangerous conflict between humans and wildlife, managing and mitigating human-elephant conflict (HEC) appears at times to be a political and logistic impossibility. This conservation flagship species (e.g. as used by the Wildlife Conservation Society or the World Wildlife Federation) can also act as a political flagship for a generalized desire to mitigate conflict by eradicating wildlife (which until recently existed to a large extent outside of protected areas in Eastern Africa, but sadly, no longer [Caro and Scholte 2007; Ogutu et al. 2011]), and to allow unchecked and unplanned agricultural expansion into areas formerly used by elephants but unused by people. Finally, livelihood threats from elephants can act as a metaphor for inequality and lack of control over land tenure and revenue sharing between protected areas and local inhabitants. Issues of disenfranchisement from land titles and from revenue-sharing were exemplified by the August 2012 spearing of elephants and lions by disaffected Maasai in Amboseli National Park, Kenya. Proposals for revenue-sharing between the National Park authorities (Kenya Wildlife Service) and local residents were originally made in 1992 (see Kangwana 1993) but not implemented fully by 2012 (see also Groom and Harris 2008). Additional pre-election politicians' promises about transfer of land ownership to communities appear to have exacerbated local discontent; the end result was a devastating and deadly spearing campaign against wildlife similar to that seen in the 1970s when the National Park was originally gazetted (see Lindsay 1987).

Do elephants cause more livelihood losses than birds, rodents or insects, or are these animals along with their damage simply more visible due to their large size, potential peril to humans and capacity for large-scale crop depredations (Hill 2005)? Elephant damage to crops can range from a few footprints and uprooted plants to an entire crop devastated after the passage of a large female group (Gunn 2009; Kiiru 2012). Insects, drought and even domestic animals are commonly the most significant problems for reductions in crop yields (Webber 2006), while pigs and primates often cause the most substantial and persistent losses after the smaller 'vermin' (e.g. figure 1.2; rodents or flocks of quelea); pigs are particularly difficult to control being nocturnal, smaller and therefore less visible (Priston 2009) (table 1.3). Rankings of crop pests where elephants and primates were both present find elephants, primates and pigs in the top 10, but where predators and livestock come into contact, then predators were increasingly perceived as the major problem animals over crop-raiders (see Holmern, Nyahongo and Røskaft 2007). Economic and cultural values both play central roles in perceptions of the severity of a pest or problem species. Throughout Asia and Africa, elephants also cause loss of life for people (Sukumar 1993; Kangwana 1995; Hoare 2000), although the risks are hard to quantify for humans who encounter elephants. Elephant encounters are probably far less risky than travelling in a motor vehicle on any developing economy road and people often hold detailed knowledge and judgements of the relative risks that elephants do pose (Sitati and Ipara 2012). Furthermore, in risk mapping elephants emerge as relatively low in overall perceived threats (table 1.2B). But the perception is one of quite legitimate personal fear, as well as a more generalized fear fostered by constant negative publicity.

Table 1.3. Perceptions of pests in East African areas with primates plus elephants present and where similar methods for assessing pests were used. Respondents were asked to rank what species they thought their top crop pests were, and a summed rank index was derived from these rankings (see Gillingham 1998). Rank 1 = most highly ranked, NP = not present to be ranked, blank = not mentioned.

Species	Species Rank by Locality		
	Kibale, Uganda[1] (n = 15 focus groups)	Mgeta River, Tanzania[2] (n = 198 households)	Laikipia, Kenya[3] (n = 342 respondents)
Baboon (*Papio* sp.)	1	4	4
Elephant (*Loxodonta africana*)	2	7	1
Redtailed monkey (*Cercopithecus ascanius*)	3	NP	NP
Bush pig (*Potomochoerus porcus*)	4	1	8
Vervet monkey (*Chlorocebus aethiops*)	5	2	6
Colobus monkey (*Colobus* sp.)	6	NP	NP
Chimpanzee (*Pan troglodytes*)	7	NP	NP
Leopard (*Panthera pardus*)			2
Hyaena (*Crocuta crocuta*)			3
Lion (*Panthera leo*)	NP		5
Porcupine (*Atherurus africanus*)			7
Rats (*Rattus* sp.)		3	
Buffalo (*Syncerus caffer*)		5	
Hippopotamus (*Hippopotamus amphibious*)	NP	6	
Other (insects, birds, cane rats)		8	9

1. MacKenzie and Ahabonya 2012, forest edge, livestock excluded, open ended discussion
2. Gillingham and Lee 2003, savannah agriculturalists, open response question
3. Graham 2007, savannah mixed ranchers, small agriculturalists, pastoralists, open response question

The term 'human-elephant conflict' (or HEC) arose in the mid-1990s as elephant populations began to increase slightly after the introduction of the CITES ban on the trade in ivory in 1989 (Kangwana 1995). The 'elephant problem' (Glover 1963; Laws, Parker and Johnstone 1975) – the all-too-human perception of 'too many large mammals in the wrong place at the wrong time, and eating too many trees' – had previously been a common metaphor for the justification of the elimination of elephants from areas of human use or control. This metaphor was gradually replaced by HEC as both human and elephant populations expanded and increasingly encountered each other outside protected areas.

HEC has now come to dominate popular perceptions of elephants in many developing country or conservation management contexts, while at the same time the approaching extinction of African forest elephants (Maisels et al. 2013) and the vast reduction in numbers of savannah elephants due to a renewed and aggressive global trade in ivory is ongoing (Wasser et al. 2010). Meanwhile, disproportionate complaints of crop losses due to elephants and protests by local farmers represent a 'weapon of the weak' (Scott 1985) in a struggle to redress social imbalances in the nature of the costs incurred and any benefits gained from the continued presence of elephants. The livelihood risks articulated in Laikipia were most associated with erratic rainfall and drought, which resulted in famine and disease – also high in Guinea-Bissau (table 1.2). We have as yet insufficient livelihood risk maps to explore more generally the perceived livelihood consequences of living with elephants, but in order to understand the risks posed by elephants in a livelihood as opposed to a crop-raiding context, these are urgently needed.

Given the hostility and real dangers to people generated by elephants that share human space, customs and aesthetic appreciation (Kuriyan 2002), religious values or traditional use of elephants (for ceremonies, e.g. Sri Lanka, or timber work, e.g. Burma) will not sustain populations of elephants in localities where they coexist with small-scale agriculture or even large plantations that require a human workforce. Even among traditionally tolerant people in India, elephants have become the 'Wicked Problem' (Bal et al. 2011), and their continued existence is now globally imperilled.

Many suggestions have been made for managing the problem of HEC. The first of these is to identify 'problem' individuals and then 'remove' these troublemakers. Genetic and observational studies suggest that most individuals, or at least most adult males, within a population will crop-raid whenever they are in proximity to attractive crops (Graham et al. 2009, 2010; Chiyo et al. 2011). There are opportunities for any elephant to become a raider, including females in family groups (Sitati and Ipara 2012). Camera traps and GSM signalling collars can help to identify persistent local raiders and warn farmers of their approach (Graham and Ochieng 2008; Graham, Adams and Kahiro 2012; chapter 10, this volume), but promoting the concept of the problem animal whose removal will speedily halt the problem is irresponsible. Removal merely creates opportunities for others to move in and exploit a rich and attractive resource. While removal may be seen as an easy and direct management strategy – and from the perspective of the afflicted farmers, at least 'something is being done' about a problem – removal followed by the invasion of replacement raiders may create even more hostility to elephants (and the ineffective managers). The management strategy of problem-animal removal (usually a euphemism for elimination) has been applied to many species of animals other than elephants (see Strum 2010) and resulted in landscapes such as those in the United Kingdom devoid of any mid- to large-sized wildlife. As suggested in figure 1.2, classification of a species or of particular types of a species (males, raiders) as 'vermin' allows for their removal without recourse to ethics or considerations of conservation. In the hierarchy of constructed beliefs about wildlife, a crop-raider becomes legitimized as an enemy.

A further question is can the behaviour of crop-raiders be managed? The cognitive capacities of elephants in relation to crop-raiding have not often been assessed, but we need to take these into account when planning management strategies (Webber et al. 2011). Males have specific 'hide and emerge when safe' cognitive strategies to facilitate human-free crop-raiding (Graham et al. 2009; Chiyo et al. 2011). We still, even after fifteen to twenty years of research, know relatively little about the selection of crops, or their importance in the ecology of elephants, which makes it more difficult to determine vulnerability and to prevent or minimize elephant incursions locally. For example, siting of villages or farms near water habitually used by elephants merely adds to the risks of raiding or conflict (Gunn 2009). There are a few sites where elephant-proof fences have acted as deterrence, while chilli, bee hives and other early warning devices can help people manage the elephants when they attempt to crop-raid (Osborn and Parker 2003; Sitati and Walpole 2006; King et al. 2009). The use of bee hives in combination with fences can be very effective as a deterrence (King et al. 2009), but when elephants damage hives, this can also impact on livelihoods and honey income. Typically elephant-proof fences are expensive and require power, almost constant maintenance and most importantly a communal effort for their construction and repair. Elephants can also learn how to climb over or break fences (even assisting each other). A further consideration about fences is that these may merely create the illusion of action and protection (Evans and Adams 2016), and when elephants violate this illusion, perceptions of the species may become even more negative than those where people are exposed to regular raiding and can use their knowledge to manage their own risks.

At this point in time, either elephants will be eliminated or we will need to persuade people to tolerate and coexist with elephants. The extinction of elephants is proceeding apace, and HEC is exacerbating this process. Do elephants have a right to exist alongside humans? Or are we more willing to prioritize the livelihoods of our own species over those of non-humans (see discussions in Hill 2002; McLennan and Hill 2012)? Ethically, at least to some of us (Poole et al. 2011), this latter attitude appears untenable. However, how can we promote tolerance? When people can predict the presence of elephants and therefore avoid them, when they can be warned and have specific actions that will help deter elephants, when they have a communal interest in allowing elephants into their areas with few or well understood risks to themselves, then perhaps we can create tolerance. These contexts are, unfortunately for elephants, extremely rare.

Conclusion

These case studies illustrate some of the issues associated with how local people perceive of community-based conservation initiatives and, more significantly, with how the political context of conflict with wildlife is defined and experienced. First, define your community – none of these are the same, none of the social, political or historical experiences will be common across our assessment contexts (Gillingham 1997; Nelson 2000; Spiteri and Nepal 2006). Defining a community – which is essential in order to construct social capital and communal actions – remains difficult at a local level, both for conservation practitioners and for the local people themselves. Communities bordering protected areas are also in the process of change – as noted for Guinea-Bissau, the communication revolution has impacted on livelihoods and therefore on attitudes to constraints on livelihoods due to wildlife or protected-area management.

Secondly, for effective action at local levels, communities need a set of shared values and expectations and a common identity. As discussed, when communities lack social cohesion and social capital, individuals are less likely to feel enfranchised or participatory. Power, wealth, status differentials, religious beliefs and the economic and social roles of women and girls all contribute to a lack of common attitudes. Environmental actors cannot make assumptions across individuals about shared core values or the willingness to accept communal or individual costs resulting from biodiversity protection. Communal conservation actions may be most effective when the full diversity or heterogeneity of a community is recognized and included in planning and actions, as specified by Arian Spiteri and Sanjay Nepal (2006). Beliefs and perceptions that can conflict with conservation outcomes need to be explored prior to program implementation. As has been shown in a number of studies, a distorted perception of conflict arises based on infrequent events combined with a sense of continued vulnerability (Priston 2005; Inskip and Zimmermann 2009). Meanwhile, positive perceptions can potentially be harnessed for conservation engagement (Gadgil, Berkes and Folke 1993; Ferraro and Kramer 1997; Kuriyan 2002; Nyhus et al. 2005) even when associated with personal costs, but we as practitioners need to know how to achieve this. Existing social and aesthetic drivers can be harnessed to change perception and actions towards pests or protected areas, even given the obvious need for livelihood empowerment close to protected areas.

Are there new ways to seek a greater understanding of what conflict means in psychological terms? While we can quantify livelihood risks and opportunity and economic costs, we only infrequently address the perceptual drivers of conflict. Are there new techniques that we can use as practitioners to assess how people perceive of and value wildlife and apply these to overturning negative perceptions? Assessing people's ecological knowledge of species of conservation concern (e.g. Gadgil, Berkes and Folke 1993; Kellert 1993; Becker and Ghimire 2003; Kaltenborn et al. 2006; Borgerhoff Mulder et al. 2009; Ross, Vreeman and Lonsdorf 2011; Sitati and Ipara 2012) may provide much needed deeper insights into when and why negative perceptions arise and therefore suggest the means to change these and thus manage conflict and create tolerance.

Thirdly, management of conflict and biodiversity protection will only come about if there is a locally based communal willingness to see tolerance emerge, and where biodiversity has a recognized and shared value among members of the community. Giving voice to elements of biodiversity, enabling this voice in stakeholder dialogues, may also provide a mechanism for a more equal spread of social and ecological justice. And we need to avoid contexts where mitigation action by practitioners simply sustains perceptions of conflict, setting up unrealizable expectations (see Groom and Harris 2008). Other questions for those of us seeking to make tolerance come about are how do we fund and sustain emergent local activities, and what is the role of local education? Empowering people to be part of their own narratives of conservation ('citizen science', e.g. Irwin 1995) can play a significant role in enabling conservation actions (for example, Community-based Biodiversity Conservation Films [www.conservationfilms.org]).

Elephants exemplify the problems we face in attempting community conservation initiatives in politicized, media-dominated and globalized economies. That elephants are approaching local extirpation is unsurprising given the demands for ivory and the lack of compliance with protective legislation. Demonizing elephants (e.g. Jadhav and Barua 2012) gives people the psychological permission to exploit or eradicate them; they are pests after all and 'a wicked problem', so if their removal generates income and eliminates a risk to people, then all the better. Greed and economics drive the trade in ivory, but allowing this trade to happen, being complicit in the killing of a protected species for its market value, is made all the easier when the species is an unmanageable five-ton 'pest'.

We also need to review the terms we use in conservation practice, since language is as important as action and, indeed, empowers action. As discussed earlier, we need to be aware of the root causes of perceptions of pests, as well as the local political context in which these pests operate. We urgently need to act at the grassroots level to assess local knowledge of the reality of animal behaviour as well as the perceptions of a pest. As I have said elsewhere, when humans see themselves as being in conflict with animals, the animals simply will not and cannot win (Lee 2010). How, if at all, can we change this dynamic so that both humans and animals can win through some form of coexistence and tolerance? This is the key question to sustaining people and wildlife together over the next century.

Acknowledgements

I thank all my collaborators and students over many years for their input, novelty of approaches and the generosity of their sharing of ideas: the Amboseli Elephant Research Project and the people of Amboseli for engagement with HEC issues; Sarah Gillingham, and the patient people in the villages around the Selous Game Reserve; Max Graham for sharing his innovative work in Laikipia, Kenya; Nancy Priston for actually measuring crop losses and trying to devise solutions on Buton Island, Indonesia, and for inviting me to participate in this meeting; the Mikumi Darwin project in Tanzania (Jody Gunn, Dawn Hawkins and Guy Norton from Anglia Ruskin University); Jo Woodman for collaborations in her work at Pench Tiger Reserve, India. Thanks also to Kate Nowak (Zanzibar and HEC in the Udzungwas, Tanzania), John Fanshawe (Birdlife), Keith Lindsay (EDG, Oxford), Jo Setchell (Durham), Bill Adams (Downing College, Cambridge) and Kate Hill and Amanda Webber (Oxford Brookes) for comments, suggestions, debates and purloined ideas.

Phyllis Lee is a Professor at the University of Stirling, Scotland. She researches evolutionary ecology, conservation and human-animal interactions. She has published extensively on elephants and primates, as well as issues in community conservation.

Note

1. I use the phrase 'so-called' here to emphasize the point that it is often unclear who has designated species as 'problem' animals or why.

References

Arluke, A., and C. Sanders. 1996. *Regarding Animals*. Philadelphia, PA: Temple University Press.

Bal, P., et al. 2011. 'Elephants Also Like Coffee: Trends and Drivers of Human–Elephant Conflicts in Coffee Agroforestry Landscapes of Kodagu, Western Ghats, India', *Environmental Management* 48: 263–275.

Bandara, R., and C. Tisdell. 2004. 'The Net Benefit of Saving the Asian Elephant: A Policy and Contingent Valuation Study', *Ecological Economics* 48: 93–107.

Baral, N., and J.T. Heinen. 2008. 'Resources Use, Conservation Attitudes, Management Intervention and Park-People Relations in the Western Terai Landscape of Nepal', *Environmental Conservation* 34: 64–72.

Baron, J., and M. Spranca. 1997. 'Protected Values', *Organizational Behavior and Human Decision Processes* 70: 1–16.

Barros, F.B., H.M. Pereira and L. Vincente. 2011. 'Use and Knowledge of the Razor-Billed Curassow *Pauxi tuberosa* (Spix, 1825) (Galliformes, Cracidae) by a Riverine Community of the Oriental Amazonia, Brazil', *Journal of Ethnobiology and Ethnomedicine* 7(1). doi:10.1186/1746-4269-7-1.

Becker, C.D., and K. Ghimire. 2003. 'Synergy between Traditional Ecological Knowledge and Conservation Science Supports Forest Preservation in Ecuador', *Conservation Ecology* 8(1): 1.

Bennett, E.L. 2011. 'Another Inconvenient Truth: The Failure of Enforcement Systems to Save Charismatic Species', *Oryx* 45: 476–479.

Borgerhoff Mulder, M., et al. 2009. 'Knowledge and Attitudes of Children of the Rupununi: Implications for Conservation in Guyana', *Biological Conservation* 142: 879–887.

Bowles, S., and H. Gintis. 2004. 'The Evolution of Strong Reciprocity: Cooperation in Heterogeneous Populations', *Theoretical Population Biology* 65: 17–28.

Browne-Nuñez, C., and S.A. Jonker. 2008. 'Attitudes toward Wildlife and Conservation across Africa: A Review of Survey Research', *Human Dimensions of Wildlife* 13: 49–72.

Caro, T.M., and P. Scholte. 2007. 'When Protection Falters', *African Journal of Ecology* 45: 233–235.

Chiyo, P.I., et al. 2011. 'Using Molecular and Observational Techniques to Estimate the Number and Raiding Patterns of Crop-Raiding Elephants', *Journal of Applied Ecology* 48: 788–796.

Costa, S. 2011. 'Social Perceptions of Non-Humans in Tombali (Guinea-Bissau, West Africa): A Contribution to Chimpanzee (*Pan troglodytes verus*) Conservation', PhD thesis, University of Stirling.

Costa, S., et al. 2008. 'Social Perception of Non-Humans in Tombali (Guinea Bissau, West Africa): An Anthropological Contribution to Chimpanzee Conservation', *Folia Primatologica* 79: 290–291.

Evans, L.A. and Adams, W.M., 2016. 'Fencing Elephants: The Hidden Politics of Wildlife Fencing in Laikipia, Kenya', *Land Use Policy* 51: 215–228.

Ferraro, P.J., and A. Kiss. 2002. 'Direct Payments to Conserve Biodiversity', *Science* 298: 1718–1719.

Ferraro, P.J., and R.A. Kramer. 1997. 'Compensation and Economic Incentives: Reducing Pressure on Protected Areas', in R. Kramer and C. van Schaik (eds), *Last Stand: Protected Areas and the Defense of Tropical Biodiversity*. New York: Oxford University Press, pp. 187–211.

Flack, J.C., F.M.B. de Waal and D.C. Krakauer. 2005. 'Social Structure, Robustness, and Policing Cost in a Cognitively Sophisticated Species', *The American Naturalist* 165: E126–E139.

Flinton, F. 2003. *Women, Gender and ICDPs in Africa: Lessons Learnt and Experiences Shared*. London: International Institute for Environment and Development.

Fisher, M. 2016. 'Whose Conflict is it Anyway? Mobilizing Research to Save Lives', *Oryx* 50: 377–378.

Fuentes, A., and L.D. Wolfe. 2002. *Primates Face to Face: The Conservation Implications of Human-Nonhuman Primate Interconnections*. Cambridge: Cambridge University Press.

Gadd, M.E. 2005. 'Conservation Outside of Parks: Attitudes of Local People in Laikipia, Kenya', *Environmental Conservation* 32: 50–63.

Gadgil, M., F. Berkes and C. Folke. 1993. 'Indigenous Knowledge for Biodiversity Conservation', *Ambio* 22: 151–156.

Giddens, A., M. Duneier and R.P. Appelbaum. 2003. *Introduction to Sociology*. New York: W.W. Norton and Co.

Gillingham, S.G. 1997. 'Do All Peasant Farmers Look Alike? The Socioeconomic Context for Community Wildlife Management around the Selous Game Reserve, Tanzania', SCP Discussion Paper No. 22, Dar es Salaam, Selous Conservation Programme.

———. 1998. 'Giving Wildlife a Value: A Case Study of Community Wildlife Management around the Selous Game Reserve, Tanzania', PhD thesis, University of Cambridge.

Gillingham, S.G., and P.C. Lee. 1999. 'The Impact of Wildlife-Related Benefits on the Conservation Attitudes of Local People around Selous Game Reserve, Tanzania', *Environmental Conservation* 26: 218–228.

———. 2003. 'People and Protected Areas: A Study of Local Perceptions of Wildlife Crop-Damage Conflict in an Area Bordering the Selous Game Reserve, Tanzania', *Oryx* 37: 316–325.

Glover, J. 1963. 'The Elephant Problem at Tsavo', *African Journal of Ecology* 1: 30–39.

Graham, M.D. 2007. 'Coexistence in a Land Use Mosaic? Land Use Risk and Elephant Ecology in Laikipia District, Kenya', PhD thesis, University of Cambridge.

Graham, M.D., W.M. Adams and G.N. Kahiro. 2012. 'Mobile Phone Communication in Effective Human–Elephant Conflict Management in Laikipia County, Kenya', *Oryx* 46: 137–144.

Graham, M.D., et al. 2009. 'The Movement of African Elephants in a Human Dominated Land Use Mosaic', *Animal Conservation* 12: 445–455.

———. 2010. 'Patterns of Crop-Raiding by Elephants, *Loxodonta africana*, in Laikipia, Kenya, and the Management of Human–Elephant Conflict', *Systematics and Biodiversity* 8: 435–445.

Graham, M.D., and T. Ochieng. 2008. 'Uptake and Performance of Farm-Based Measures for Reducing Crop Raiding by Elephants *Loxodonta africana* among Smallholder Farms in Laikipia District, Kenya', *Oryx* 42: 76–82.

Gregory, R., S. Lichtenstein and P. Slovic. 1993. 'Valuing Environmental Resources: a Constructive Approach', *Journal of Risk and Uncertainty* 7: 177–197.

Groom, R., and S. Harris. 2008. 'Conservation on Community Lands: The Importance of Equitable Revenue Sharing', *Environmental Conservation* 35: 242–251.

Gunn, J. 2009. 'Human Elephant Conflict around Mikumi National Park, Tanzania: A Quantitative Evaluation', PhD thesis, Anglia Ruskin University, Cambridge.

Haenn, N. 1999. 'The Power of Environmental Knowledge: Ethnoecology and Environmental Conflicts in Mexican Conservation', *Human Ecology* 27: 477–491.

Hill, C.M. 1998. 'Conflicting Attitudes towards Elephants around the Budongo Forest Reserve, Uganda', *Environmental Conservation* 25: 244–250.

———. 2002. 'Primate Conservation and Local Communities: Ethical Issues and Debates', *American Anthropologist* 104: 1184–1194.

———. 2005. 'People, Crops and Primates: A Conflict of Interests', in J.D. Paterson and J. Wallis (eds), *Commensalism and Conflict: The Human–Primate Interface*. Norman, OK: American Society of Primatologists, pp. 40–59.

Hoare, R. 2000. 'African Elephants and Humans in Conflict: the Outlook for Co-Existence', *Oryx* 34: 34–38.

Hockings, K.J., and C. Sousa. 2011. 'Human-Chimpanzee Sympatry and Interactions in Cantanhez National Park, Guinea-Bissau: Current Research and Future Directions', *Primate Conservation* 26(1): 57–65.

Holmern, T., J. Nyahongo and E. Røskaft. 2007. 'Livestock Loss Caused by Predators Outside the Serengeti National Park, Tanzania', *Biological Conservation* 135: 518–526.

Humphrey, N. 2007. 'Society of Selves', *Philosophical Transactions of the Royal Society*, B, 362: 745–754.

Igoe, J., and D. Brockington. 2007. 'Neoliberal Conservation: a Brief Introduction', *Conservation and Society* 5: 432–449.

Inskip, C., and A. Zimmermann. 2009. 'Human-Felid Conflict: A Review of Patterns and Priorities Worldwide', *Oryx* 43: 18–34.

Irwin, A. 1995. *Citizen Science: A Study of People, Expertise, and Sustainable Development*. London: Routledge.

Jadhav, S., and M. Barua. 2012. 'The Elephant Vanishes: Impact of Human–Elephant Conflict on People's Wellbeing', *Health and Place* 18: 1356–1365.

Kaltenborn, B.P., et al. 2006. 'Animal Preferences and Acceptability of Wildlife Management Actions around Serengeti National Park, Tanzania', *Biodiversity and Conservation* 15: 4633–4649.

Kangwana, K.F. 1993. 'Conservation and Conflict: Elephant-Maasai Interaction in Amboseli Kenya', PhD thesis, University of Cambridge.

———. 1995. 'Human–Elephant Conflict: The Challenge Ahead', *Pachyderm* 19: 11–14.

Karanth, K.K., et al. 2008. 'Examining Conservation Attitudes, Perspectives and Challenges in India', *Biological Conservation* 141: 2357–2367.

Kellert, S.R. 1993. 'Values and Perceptions of Invertebrates', *Conservation Biology* 7: 845–855.

———. 1996. *The Value of Life: Biological Diversity and Human Society*. Washington, DC: Island Press.

———. 2008. 'A Biocultural Basis for an Ethic towards the Natural Environment', in L. Rockwood, R. Stewart and T. Dietz (eds), *Foundations of Environmental Sustainability: The Co-Evolution of Science and Policy*. Oxford: Oxford University Press.

Kiiru, W. 2012. 'Understanding the Spatial, Temporal and Socio-Economic Factors Affecting Human-Elephant Conflict around Amboseli National Park, Kenya', PhD thesis, University of Kent, Canterbury.

King, L.E., et al. 2009. 'Beehive Fence Deters Crop-Raiding Elephants', *African Journal of Ecology* 47: 131–137.

Kuriyan, R. 2002. 'Linking Local Perceptions of Elephants and Conservation: Samburu Pastoralists in Northern Kenya', *Society and Natural Resources* 15: 949–957.

Laws, R.M., I.C.S. Parker and R.C.B. Johnstone. 1975. *Elephants and their Habitats: The Ecology of Elephants in North Bunyoro, Uganda*. Oxford: Clarendon Press.

Lee, P.C. 2010. 'Sharing Space: Can Ethnoprimatology Contribute to the Survival of Non-Human Primates in Human-Dominated Globalized Landscapes?' *American Journal of Primatology* 71: 925–931.

Lee, P.C., and M.D. Graham. 2006. 'African Elephants (*Loxodonta africana*) and Human–Elephant Interactions: Implications for Conservation', *International Zoo Yearbook* 40: 9–19.

Lindsay, W.K. 1987. 'Integrating Parks and Pastoralists: Some Lessons from Amboseli', in D. Anderson and R. Grove (eds), *Conservation in Africa*. Cambridge: Cambridge University Press, pp. 149–168.

Lindsey, P.A., J.T. du Toit and M.G.L. Mills. 2005. 'Attitudes of Ranchers towards African Wild Dogs *Lycaon pictus*: Conservation Implications on Private Land', *Biological Conservation* 125: 113–121.

MacKenzie, C.A., and P. Ahabonya. 2012. 'Elephants in the Garden: Financial and Social Costs of Crop Raiding', *Environmental Economics* 75: 72–82.

Madden, F. 2004. 'Creating Coexistence between Humans and Wildlife: Global Perspectives on Local Efforts to Address Human-Wildlife Conflict', *Human Dimensions of Wildlife* 9: 247–257.

Maisels, F., et al. 2013. 'Devastating Decline of Forest Elephants in Central Africa', *PLoS ONE* 8: e59469. doi:10.1371/journal.pone.0059469.

McLennan, M.R., and C.M. Hill. 2012. 'Troublesome Neighbours: Changing Attitudes towards Chimpanzees (*Pan troglodytes*) in a Human-Dominated Landscape in Uganda', *Journal for Nature Conservation* 20: 219–227.

Mehta, J.N., and J.T. Heinen. 2001. 'Does Community-Based Conservation Shape Favorable Attitudes among Locals? An Empirical Study from Nepal', *Environmental Management* 28: 165–177.

Murphree, M.W. 1993. *Communities as Resource Management Institutions*. Gatekeeper Series, no. 36. London: International Institute for Environment and Development.

Naidoo, R., and T.H. Ricketts. 2006. 'Mapping the Economic Costs and Benefits of Conservation', *PLoS Biology* 4(11): e360. doi:10.1371/journal.pbio.0040360.

Nelson, F. 2000. 'Sustainable Development and Wildlife Conservation in Tanzanian Maasailand', *Environment, Development and Sustainability* 2: 107–117.

Newton, K. 2001. 'Trust, Social Capital, Civil Society, and Democracy', *International Political Science Review* 22: 201–214.

Nyhus, P.J., et al. 2005. 'Bearing the Costs of Human-Wildlife Conflict: The Challenges of Compensation Schemes', in R. Woodroffe, S. Thirgood and A. Rabinowitz (eds), *People and Wildlife: Conflict or Coexistence*. Cambridge: Cambridge University Press, pp. 107–122.

Ogutu, J.O., et al. 2011. 'Continuing Wildlife Population Declines and Range Contraction in the Mara Region of Kenya during 1977–2009', *Journal of Zoology* 285: 99–109.

Osborn, F.V., and G. Parker. 2003. 'Towards an Integrated Approach for Reducing the Conflict between Elephants and People: A Review of Current Research', *Oryx* 37: 80–84.

Peterson, M.N., et al. 2010. 'Rearticulating the Myth of Human–Wildlife Conflict', *Conservation Letters* 3: 74–82.

Poole, J.H., et al. 2011. 'Ethical Approaches to Elephant Conservation', in C.J. Moss, H. Croze and P.C. Lee (eds), *The Amboseli Elephant: A Long-term Perspective on a Long-Lived Mammal*. Chicago: University of Chicago Press, pp. 318–326.

Priston, N.E.C. 2005. 'Crop Raiding by *Macaca ochreata brunnescens* in Sulawesi: Reality, Perceptions and Outcomes for Conservation', PhD thesis, University of Cambridge.

———. 2009. 'Exclosure Plots as a Mechanism for Quantifying Damage to Crops by Primates', *International Journal of Pest Management* 55: 243–249.

Riley, E.P. 2006. 'Ethnoprimatology: Toward Reconciliation of Biological and Cultural Anthropology', *Ecological and Environmental Anthropology* 2: 75–86.

Ross, S.R., V.M. Vreeman and E.V. Lonsdorf. 2011. 'Specific Image Characteristics Influence Attitudes about Chimpanzee Conservation and Use as Pets', *PLoS ONE* 6(7): e22050. doi:10.1371/journal.pone.0022050.

Sagoff, M. 1998. 'Aggregation and Deliberation in Valuing Environmental Public Goods: A Look Beyond Contingent Pricing', *Ecological Economics* 24: 213–230.

Schultz, P.W., et al. 2005. 'Values and Their Relationship to Environmental Concern and Conservation Behavior', *Journal of Cross-Cultural Psychology* 36: 457–475.

Scott, J.C. 1985. *Weapon of the Weak: Everyday Forms of Peasant Resistance*. New Haven, CT: Yale University Press.

Sitati, N.W., and H. Ipara. 2012. 'Indigenous Ecological Knowledge of a Human-Elephant Interaction in Transmara District, Kenya: Implications for Research and Management', *Advances in Anthropology* 2: 107–111.

Sitati, N.W., and M.J Walpole. 2006. 'Assessing Farm-Based Measures for Mitigating Human-Elephant Conflict in Transmara District, Kenya', *Oryx* 40: 279–286.

Smith, K., C.B. Barrett and P.W. Box. 2000. 'Participatory Risk Mapping for Targeting Research and Assistance: With an Example from East African Pastoralists', *World Development* 28: 1946–1959.

Soltis, J., R. Boyd and P.J. Richerson. 1995. 'Can Group-Functional Behaviors Evolve by Cultural Group Selection? An Empirical Test', *Current Anthropology* 36: 473–494.

Spiteri, A., and S.K. Nepal. 2006. 'Incentive-Based Conservation Programs in Developing Countries: A Review of Some Key Issues and Suggestions for Improvements', *Environmental Management* 37: 1–14.

Strum, S.C. 2010. 'The Development of Primate Raiding: Implications for Management and Conservation', *International Journal of Primatology* 31: 133–156.

Sukumar, R. 1993. *The Asian Elephant: Ecology and Management*. Cambridge: Cambridge University Press.

Swift, M.J., M.N. Izac and M. van Noordwijk. 2004. 'Biodiversity and Ecosystem Services in Agricultural Landscapes: Are we Asking the Right Questions?' *Agriculture, Ecosystems and Environment* 104: 113–134.

Tomasello, M., et al. 2005. 'Understanding and Sharing Intentions: The Origins of Cultural Cognition', *Behavioral and Brain Sciences* 28: 675–735.

UNODC. 2007. *Cocaine Trafficking in West Africa: The Threat to Stability and Development (with special reference to Guinea-Bissau)*. United Nations Office on Drugs and Crime.

Vira, B., and W.M. Adams. 2009. 'Ecosystem Services and Conservation Strategy: Beware the Silver Bullet', *Conservation Letters* 20: 1–5.

Wasser, S., et al. 2010. 'Elephants, Ivory and Trade', *Science* 327: 1331–1332.

Webber, A.D. 2006. 'Primate Crop Raiding in Uganda: Actual and Perceived Risks around Budongo Forest Reserve', PhD Thesis, Oxford Brookes University.

Webber, C.E., et al. 2011. 'Elephant Crop-Raiding and Human-Elephant Conflict in Cambodia', *Oryx* 45: 243–251.

Zimmerman, B., et al. 2001. 'Conservation and Development Alliances with the Kayapó of South-Eastern Amazonia, a Tropical Forest Indigenous People', *Environmental Conservation* 28: 10–22.

2
Block, Push or Pull?
Three Responses to Monkey Crop-Raiding in Japan

John Knight

The Japanese macaque, *Macaca fuscata*, is a major crop pest in rural Japan. This chapter is about how people deal with this problem. I focus on three responses to monkey crop-raiding: obstructing monkey entry into villages and fields (what I call the 'block' response), chasing the monkeys back to the forest (what I call the 'push' response) and luring the monkeys away from fields using food handouts at another site (what I call the 'pull' response). I describe the effectiveness of these interventions and monkey resistance to them. I then show how these responses lead to a change in the perception of monkey crop-raiding, which ceases to appear simply as a monkey problem and comes to be seen as something caused by other people. Monkey crop-raiding in Japan becomes a site of social division among humans as much as a site of conflict between humans and monkeys.

My theoretical point of reference in this discussion is monkey mobility[1] and its determinants. Food exploitation among non-human primates is informed by a concern with safety vis-à-vis potential predators (Miller 2002). The mobility and use of space of non-human primates can therefore be accounted for in terms of both their attraction to food and their avoidance of threat, as well as the trade-off between them. This general point applies to situations in which the range of non-human primates overlaps with human space. In these circumstances, human threat and human food are likely to play their part in influencing primate mobility, alongside other threat and food sources. In what follows I shall examine how human threat and human food come to shape the pattern of mobility of Japan's pest monkeys.

Engai

What is known as *engai* or 'monkey damage' is widespread in regional Japan, affecting rural communities in monkey range areas across the country. In 2007 monkeys reportedly caused some ¥1.6 billion (approximately $13 million) worth of crop damage in Japan, affecting around 3,700 hectares of farmland (Ministry of Agriculture n.d.). This may well be an underestimate as much low-level monkey crop feeding tends to go unreported. The size of the crop-raiding monkey population in Japan is large and growing. In many prefectures a majority of monkey troops are engaged in crop-raiding (Watanabe 2007: 147). These monkeys feed on a wide range of crops, especially fruit and vegetables. Repeated monkey depredations and ruined harvests lead to highly negative views of the monkey *dorobō* or 'thieves' and can erode the will to continue farming.

The problem tends to be especially severe in depopulated mountain villages. A characteristic of the depopulated village is that there is much less human activity in it. During my fieldwork in the mountainous interior of Wakayama Prefecture, the village of Takayama was a place with a serious monkey problem. Residents claimed that the monkey problem worsened dramatically following the closure of the village primary school in the late 1980s. Their reasoning was that the monkeys found it much easier to enter the now much quieter village environment. Asari Tetsuichi, a mushroom-grower in Takayama, touched on this point in an interview: 'If there were lots of people around, they [the monkeys] would not enter. Even if there were only children about, making lots of noise, I don't think that they would come. But children have steadily decreased and there are only a few old people about. They [the monkeys] are clever, they know that. . . . It [the village] has become quiet, too quiet. . . . If you go there, and walk all the way through during the daytime, you will hardly meet anyone.'[2]

Farming and agroforestry activity is of special importance in this connection. The primatologist David Sprague refers to the existence of a 'buffer zone of human activity' around the village, whereby 'the foothills surrounding villages were once used, occupied, and defended by humans and their domesticated animals under traditional agriculture' (2002: 264). However, with depopulation and the decline of farming, this buffer zone has ceased to exist. Village farmland and the *satoyama* or 'village forest' (the belt of woodland around the village) no longer contains an active daytime human presence, which means that monkeys entering this zone are unlikely to be disturbed. As one recent newspaper report on the wildlife problem puts it, 'the people pressure at the boundary between forest and village has weakened' (Shikoku Shinbun 2008).

The reduced human presence makes the village a much easier place for monkeys to raid. To appreciate why this is, we must draw a distinction between the monkey's spatial boldness and its boldness vis-à-vis people. A monkey may be bold enough to enter a village field to feed on the crops there, but too timid to do so when there is a human about. Or again, when it is actually raiding a field, the appearance or sound of a human (or even a car in the distance) may be enough to send the monkey fleeing for cover in the nearby forest. Although bold enough to enter human space, the monkey is still intolerant of any human presence. The amount of crop damage that such a monkey causes is limited by the imperative of avoiding human encounters. This means that the more depopulated the village, the more vulnerable to monkey raids it tends to become. The significance of this is that even wary monkeys, that do not tolerate the presence of humans, are able to raid fields in sparsely populated places. In short, depopulation lowers the habituation threshold for monkey crop-raiding.

However, over time monkeys become bolder. Their increasing boldness is manifested not just in the frequency of their visits to the village, but also in their greater tolerance of human encounters. The presence of humans in and about the village is no longer the automatic deterrent to would-be monkey crop-raiders it was earlier. The monkey's flight distance, the distance at which a human presence causes monkeys to flee, contracts, which means the animal reacts only to close-up encounters. In a manual on *engai* countermeasures compiled by the primatologist Izawa Kōsei, a flight distance of 100 metres is taken as the standard; where this flight distance exists, the degree of monkey habituation is low and *engai* tends not to be a problem (Izawa and MSC 2005: 81). But where *engai* is rife this flight distance may decline to 50 metres or even disappear altogether as monkeys cease to flee even when people are nearby (ibid.).

Another expression of the greater boldness of monkeys is house-entry behaviour. Monkeys get into houses by opening or unlocking windows and doors, feed on the food they find in the kitchen (even, in some cases, reportedly opening refrigerators) and feed on the edible offerings (fruit, candies, cakes, etc.) made to family ancestors in the *butsudan* altar. One of the most disagreeable aspects of monkey entry into houses is the mess and odour they tend to leave behind them (especially when they defecate in the house). Monkeys also dislodge roof tiles, fiddle with outside television antennae (affecting television pictures) and interfere with telephone cables (occasionally causing village telephone lines to go down).

Engai in these conditions can readily be accounted for in terms of the combined effect of food and threat. Villages were once high-threat places; despite the appealing crops in the village, the presence of people (including hunters) there tended to deter monkeys from raiding. Threat avoidance prevailed over food appeal. But in the era of depopulation, villages have become low-threat places that no longer deter monkeys from entering. Today, food appeal rather than threat avoidance determines the monkey's relationship with the village.

Blocking the Monkeys

Let us now look at how villages respond to this monkey problem. Perhaps the most obvious response is to block the entry of the monkeys into the village. Fencing is a widespread response to the general *engai* problem in rural Japan. Fences are erected around individual fields, around a wider area of farmland and along the perimeter of the village (the boundary with the forest). Hard-mesh fences are used, sometimes extending horizontally at the top to form a roof above the field – a structure sometimes described as an *ori* or 'cage'. Resorting to 'cages' as a means of defending farmland from monkeys occasions wry comments from farmers. As one man in Owase in Mie Prefecture put it, 'for human beings to cultivate inside cages is a preposterous state of affairs, but as protection against damage this is the only method that works' (Chūnichi Shinbun 1997). The raiding monkeys were turning the village into a kind of human zoo!

There are other, less expensive and more easily maintainable fencing solutions available to the farmer. In recent years, flexible fencing, which is difficult for monkeys to climb over, has become popular. But simple fencing is likely to be less effective against the agile monkey than against other animals, as monkeys are able to climb over fences in a way that animal pests such as wild boar and deer cannot. This is the background to the spread of electric fencing in villages suffering from *engai*. Unlike many of the other tactics used to try and stop *engai*, electric fencing has proved quite effective and has many champions in rural localities. The primatologist Fukuda Fumio has concluded that electric fencing is the only practical long-term solution to the problem of monkey damage in most of rural Japan (Fukuda 1992: 293). That said, electric fencing is not totally reliable. Monkeys sometimes climb over an electric fence by deftly avoiding contact with the live parts of the structure. Moreover, the effectiveness of electric fencing is conditional on proper maintenance, which is not always carried out. When a fence does short out, there may be a long delay in repairing it, and where this happens repeatedly the farmer is likely to lose faith in electric fencing.

Monkey intrusion into farmland can be blocked in other ways. In Japan there is a long-established practice known as the *sarunoban* or 'monkey watch' in which somebody is stationed near fields to protect them from monkey raids (see Knight 2003: 107–109). As the monkey is a diurnal animal, *sarunoban* is in principle a day-long, dawn-to-dusk undertaking (especially as harvest time approaches). Sometimes cooperation in field-guarding involves no more than two or three households with adjacent plots of farmland, as in the Hongū village of Ōtsuga where the members of three farming households take turns patrolling their rice fields in the weeks running up to the harvest in September. But as with many other areas of village cooperation (such as festival organization), collective field-guarding is said to have greatly declined in recent decades and farming households are more likely to guard their fields themselves.

Field-guarding, performed by both men and women, is known to be hard, hot work. One elderly man described to me the practice of field-guarding as follows:

> Other people have given up guarding because it's so hard. . . . If you knew what time they were coming, that would be one thing, but you don't know, so you just go on guarding with no idea. It wouldn't be so bad if, say, you knew that they were definitely coming during the day ahead. . . . But you have to be ready to act at any time. . . . You sit there, holding on to the firecrackers [to throw at the monkeys], and you have to play the radio and the music throughout. . . . You just don't know when they will come. Just when you think they are not going to come, then they will suddenly come. . . . It's just at this sort of time [early evening] when the sun has set and it has turned dark and there is no longer a human presence [in the

fields] that they'll [finally] come! [They come] in the evening [at dusk], at dawn, or they come during the day just when you leave the fields for a while.

As these comments indicate, field-guarding is a battle of wits with the monkeys. The guard must stay alert throughout the day, making sure he avoids dozing off after lunch in the heat of the early afternoon. Where a number of fields have to be protected, the task is all the more demanding, especially if the field guard has to move about to keep all his farmland in view (such as where a field or part of a field is not in his line of sight or located in another part of the valley). Then there is the possibility the monkeys will raid the fields either before the field guard arrives or after he leaves.

Some farmers delegate the task of *sarunoban* to dogs. In Japan, dogs are known as the fiercest enemy of the monkey, an antagonism captured in the proverb, *yome to shūtome inu to saru*, 'like wife and mother-in-law, so dog and monkey'. Japanese breeds are often used for field-guarding against monkeys, but in some areas of Japan imported breeds may be deployed (for example, Shetland sheepdogs in Nagano Prefecture). However, the efficacy of posted guard dogs is said to be limited because the clever monkeys soon realize the dogs are confined and cannot harm them, and proceed to raid the fields with impunity. One option, of course, might be for the farmer to release the dog so it can chase the monkeys back into the forest. But the problem here is that dogs that chase monkeys may become so involved in the pursuit they end up getting lost in the forest.[3]

Pushing Away the Monkeys

The purpose of fencing is to block the movement of monkeys towards village fields. But there is another possibility: reform of the behaviour of the monkeys. If monkeys were to learn to keep away from villages and not raid farmland, frictions with residents could be greatly reduced.

Fear-inducement is the basis of this approach. To get monkeys to keep away from the village, villagers must make it a frightening place for them. To this end, assorted scare tactics are used against the monkeys. *Kakashi* or scarecrows are a common sight in (or at the edge of) fields, especially vulnerable forest-edge fields, and are directed against monkeys as well as other kinds of animals. Typically life-size, the *kakashi* usually has a straw-filled body, a hat of some kind (a wide-brimmed straw hat, a bonnet, a peaked cap or even a safety helmet), a face (a doll's head, a plastic noh mask or just a pair of sunglasses) and is dressed in old clothes (a smock, a shirt, a blouse, a baseball top, a raincoat and so on). The *kakashi* is usually in an upright standing position as this makes them more visible to monkeys from afar (although sometimes they are seated or slightly bent over holding a farm implement). Occasionally, the *kakashi* holds a toy gun (or a stick or rod resembling a gun) in the belief that monkeys are particularly fearful of armed hunters. Usually there is just a single *kakashi* in the field, but one sometimes sees large fields with kakashi placed on different sides.

The widespread use of *kakashi* to protect fields against monkeys would suggest that there is a belief in their efficacy as an *engai* countermeasure. However, my impression is that they are seen as more effective against animals other than monkeys. In Japan the monkey is known as an intelligent animal and one often hears farmers say that 'clever' monkeys learn to ignore the *kakashi* because they notice they are stationary and do not move around like real people. For this reason the farmer may alter the position or appearance of the *kakashi* or even contrive to have the *kakashi* move (for example, by attaching to its rear a sheet of wood that catches the wind and makes the structure sway and turn). But many farmers still consider it only a matter of time before the monkeys see through the ploy.

Other kinds of frightening imagery are used. A farmer in the Hongū village of Hoshinbo used the corpse of a trapped monkey, hung upside down near the field, to keep monkeys away from his crops (Knight 2003: 104). Alternatively, images or models of scary-looking creatures may be used. In some areas huge billboards of Godzilla – the conqueror of King Kong in the famous 1962 film *King Kong versus Godzilla* – are erected at the edge of the village facing the forest in order to protect farms (Nomoto 1994: 136). Home-made life-size models of orangutans have been used to protect fields in the belief that Japanese monkeys would be deterred by the presence of such 'monkeys' so much bigger than themselves (Nishi Nihon Shinbun Chōkan 1998). In Okazaki City in Aichi Prefecture a solar-powered animatronic tiger has been used to frighten away raiding monkeys (Suzaki 2010).

In recent years a campaign has been launched to intensify the fear factor by mobilizing villagers to behave in a more threatening way towards encroaching monkeys. The emphasis is on villagers confronting, in a noisy and aggressive way, monkeys that appear in and around the village in order to give them a fright. The idea is that, even if the monkeys have become bold enough to try and raid the fields, the frightening encounter with the villagers will henceforth dispose them to keep away.

This counterhabituation strategy is associated with a man called Inoue Masateru (Inoue 2002), an agricultural extension official working for Nara Prefecture, in collaboration with a number of primatologists, including Muroyama Yasuyuki (Muroyama 2003). Inoue argues that monkeys raid village farmland not just because there is food available there, but because they encounter little resistance when doing so and lose their erstwhile fear of humans and human space. Under the conditions of rural depopulation, the monkeys in the surrounding forest become bolder in their behaviour and prepared to raid the villages they once avoided. Monkeys do not necessarily flee when the farmer appears; while in some cases monkeys (especially adult males) aggressively approach the people they encounter.

In this situation, one way in which the village can start to make monkeys fearful of humans is by what is known as *oiharai* (or *oiage*) or 'monkey chasing'. The practice of *sarunoban* was described earlier, and superficially this may resemble *oiharai*. But *sarunoban* is a somewhat passive measure, aimed at blocking monkey access to crops. *Oiharai*, on the other hand, goes further and entails the aggressive pursuit of the monkeys as they flee back to the forest.

To have maximum effect, the monkey-chasing tactic should deploy a large number of people, use loud noisemaking accessories such as firework rockets, attack the monkeys from different directions, make the assault from high ground (as this is believed to make the monkeys feel more vulnerable) and deny the monkeys the opportunity to make an easy escape (by anticipating the likely escape path and posting people there) (Izawa and MSC 2005: 40–41). The monkey-chasers' sudden assault from all sides should throw the troop into a state of panic (ibid.: 40). Ideally, the monkeys learn from this experience that the village is an intolerably dangerous place and shift or recentre their range back towards the forest.

The scale of the human presence in depopulated villages is much less than it was, and this, in large measure, is the source of the village's vulnerability to animal intrusion. But Inoue's argument is that remaining villagers need to compensate for their diminished numerical presence by acting more assertively towards the animals they encounter in order to help restore the animals' fear of humans. The village must become a place that animals find difficult and risky to enter and, even when entered, uncomfortable and frightening to spend any time in. Inoue and his group advocate what amounts to a policy of zero tolerance, whereby villagers collectively refuse to tolerate any monkey presence in and around the village. When a monkey is sighted, villagers must instantly react with an unambiguous, unremitting hostility towards it. The aim is to impart to the monkeys the message that *ningen wa teki da* or 'humans are the enemy' (Inoue 2002: 2) and that the village is enemy territory that is best avoided.

The key point of the campaign is that fencing alone cannot be the solution to monkey encroachment. Although physically impressive, fences by themselves are unlikely to be enough to keep resourceful monkeys out of the village. What is required is a rounded response that involves not just *butsuriteki shōheki* or 'physical barriers' to monkey entry, but also *shinriteki shōheki* or 'psychological barriers' (Muroyama 2003: chap. 6). Villagers must exert a *shinritekina puresshā* or 'psychological pressure' on the monkeys that is strong enough to deter them from entering the village (Inoue 2002: 93). The best kind of fences for keeping monkeys out, in other words, are not the physical ones that can be seen, but the psychological ones that cannot be seen (ibid.: 28). Ultimately, an active rural citizenry prepared to confront monkeys encountered in and around the village is the key to reversing the advanced state of monkey habituation that underpins the present-day scale of *engai*. If *engai* monkeys meet with a consistent and uncompromising hostility from villagers, they will eventually become wary of approaching the village.

Pulling Away the Monkeys

In the previous sections of this chapter, monkey colonization of human spaces has appeared as the basic problem underlying *engai*. It followed that the solution lay in humans resisting and even reversing this advance of the monkeys. In this section I shall consider an alternative response: the exploitation rather than the reversal of monkey encroachment. Instead of pushing monkeys back to the forest, intensify the pull of the human zone, and use the appeal of the human foods to assert human control over the monkeys in a way that both exploits them as a resource and neutralizes them as pests.

One site of crop-raiding after the Second World War was at the foot of Takasakiyama. Takasakiyama is a coastal mountain that falls within the municipality of Ōita City on the northeast coast of the island of Kyushu in southwestern Japan. In the late 1940s and early 1950s, the monkeys on this mountain became serious pests, damaging crops on nearby farmland. Irate farmers tried to repel the monkeys and some demanded they be eradicated.

The local mayor responded by proposing a very different kind of solution to the problem. He suggested that City Hall should create a park on the mountain in which to feed the monkeys on a daily basis and allow them to serve as a tourist attraction. In the process, the monkeys would be diverted from farmland because monkeys that regularly feed in the park would have no need to raid the farmer's crops. The mayor claimed that his idea of feeding the Takasakiyama monkeys would 'kill two birds with one stone': it would create a tourist attraction and solve the monkey pest problem at one and the same time (Ueda Tamotsu, in Nakagawa 2003: 126).

If one thinks about it, *engai* itself is due to the attractive power of food. Villages inadvertently pull monkeys in to feed on farmland crops. When farmers block and push away monkeys, they are attempting to deny the monkeys access to the crops by means of some form of physical obstruction. But there are clearly problems with this strategy: monkeys may get through or around physical barriers or raid when the field guard has fallen asleep or gone home. The mayor's provisioning proposal offered a neat alternative: it would use the underlying cause of the crop raid – food appeal – to defeat the crop raid. Food appeal would divert the monkeys away from farmland to the feeding station.

However, provisioning, narrowly conceived, was open to the following, fairly obvious objection: that providing food crops to monkeys to protect food crops from monkeys was self-defeating and contradictory because in the end valuable human food would be lost to monkeys in any case. Recalling the above characterization of *engai* monkeys as 'thieves', one is tempted to offer the analogy of paying thieves not to steal. Yes, one stops the act of theft, but only by giving the thief what he was intent on stealing. If, in this way, one defeats the monkey thieves, this is surely a hollow victory. In other words, in its simple diversionary form, provisioning would be deficient.

This is where the second part of Ueda's idea comes in: by linking provisioning with touristic display, he could argue that providing food to monkeys to protect human food did make sense because the food given out would additionally bring in tourists willing to pay to see the monkeys attracted by it. As Mayor Ueda saw it, provisioning would not just pull monkeys away from fields, but also ultimately pull in tourists to see the monkeys. In return for the food given away to the monkeys, the human side would benefit in terms of farm protection and tourism promotion.

The mayor's monkey tourism scheme had an appealing simplicity to it. It suggested that the relationship between local people and monkeys could be transformed, almost at a stroke, from one in which monkeys undermined local livelihoods to one in which they contributed to the local economy. The linkage of provisioning with crop protection was also important politically, as it potentially broadened support for the mayor's initiative, especially among the farmers directly affected. Presented in this way, the organized feeding of the monkeys would not be a pro-tourism measure at the expense of farming, but a cross-sector initiative that would serve the interests of both tourism and farming.

The mayor implemented his idea and in spring of 1953 Takasakiyama Natural Zoo opened. The monkeys proved a big hit with the public, drawing huge crowds of visitors. More than half a million people came to the park in its first year of operation, and this annual figure steadily increased in subsequent years, so that, by its tenth year, the park was attracting almost a million and a half visitors a year (Takasakiyama 1998: 26). Takasakiyama had become (to use a term that appeared in one newspaper article on the park) a *takara no yama* or a 'treasure mountain' (Yomiuri Shinbun 1995) or (to recall another newspaper term applied to the park) a *dorubako* or 'goldmine' (literally, 'dollar box') (Ōita Gōdō Shinbun 1971). The mayor's monkey-feeding initiative had created out of nothing one of the most popular attractions in the country. Inspired by the mayor's initiative, localities in different parts of the country were soon opening their own parks – overall, forty-one monkey parks were opened (nineteen in the 1950s, seventeen in the 1960s and five in the 1970s) (see Knight 2011: 35–36).

In terms of farm protection, however, provisioning has been less successful. Regular provisioning did not instantly make *engai* a thing of the past, as reports of monkey damage in nearby settlements continued to emerge after the parks were up and running. A 1959 survey of monkey parks found that, following the onset of provisioning, monkey damage decreased in only four out of twelve parks, while in two parks the level of monkey damage had actually increased (Itani 1959: 39). More recent reports indicate that provisioning may well precede and, by implication, cause crop-raiding (Hayashi and Watanabe 2000: 86; Mito et al. 2000: 50). The precise effect of provisioning would therefore seem to vary from place to place, and while it might be an effective crop-protection measure in some localities, elsewhere it can have the opposite effect.[4]

In the long run, provisioning in response to crop-raiding proves to be counterproductive and results in monkey damage on a larger scale. Provisioned troops change their ranging behaviour and relocate away from the forest to the area around the park. This relocation of the troop range towards the park means that, even if the immediate incidence of crop-raiding declines because of the diversionary effect of the park as an alternative feeding ground, the potential for crop-raiding inevitably increases because the park tends to be part of the *satoyama* zone of the mountains that adjoins human settlements. A second reason why provisioning does not divert monkeys from farms in the long run is that it introduces monkeys to human food produce. The monkeys become accustomed not just to the food of the park, but to human-produced food more generally, including the crops grown by nearby farmers. The third reason why provisioning leads to crop-raiding is that it habituates monkeys to humans. Although unprovisioned monkeys too become habituated to humans and this is an important factor in *engai* in general, the fact remains that the extent of habituation is much greater with the provisioned monkeys of the parks.[5]

These have not been the only responses to monkey crop-raiding in postwar Japan. Another response is monkey capture and culling. In postwar Japan the monkey was delisted as a game animal, as a result of which it cannot be hunted for commercial or recreational purposes. Yet monkeys are, in effect, hunted (more precisely, trapped) in large numbers as pests. In the ten-year period 1989–1998 an average of more than six thousand monkeys were culled or removed as nuisance animals each year (Sprague 2002: 260). The figure increased to more than ten thousand a year in the late 1990s and stood at 13,145 in 2013 (Ministry of the Environment n.d.). Primatologists are concerned about this high rate of culling, which, along with habitat degradation due to the spread of timber plantations, they see as 'the major current threat to the ecological security of Japanese monkey populations' (Sprague 2002: 261). Yet culling does have a measure of support among local farmers.[6]

The Transfiguration of Engai

This chapter has described various ways in which humans interfere with the mobility of pest monkeys. But this human interference also has a larger effect, one that changes the way in which the *engai* problem is perceived. For when humans intervene in monkey mobility, they are prone to become associated with, and even held responsible for, any negative effects of that mobility. In other words, monkey mobility becomes human-associated monkey mobility – something for which one group of humans can blame another group of humans. The negative effects of human-altered monkey mobility come to be seen as human-caused instances of damage.

One example of this is where one farmer's 'block' or 'push' response has a knock-on effect on other nearby farms. By blocking out the animals through fencing or by pushing away the animals through *oiharai*, the farmer tends to deflect the problem onto neighbouring farmers. Or again where one village protects itself from *engai* through fencing and/or *oiharai*, it may end up deflecting those monkeys onto the farmland of an unprotected (or at least less well-protected) adjacent village. In both cases the party that has had the monkeys deflected onto it is forced to protect itself by following suit and erecting fencing or practicing *oiharai* itself. In this way, there develops a kind of chain reaction in response to *engai*. But in this situation the parties incurring monkey damage as a result of the protective measures taken by neighbours may well see these neighbours as responsible for the damage incurred. The actions of that farmer (or that village) caused the damage to my farm (or my village), requiring that I undertake expensive or burdensome protective measures too!

Something similar occurs with the pull strategy. We saw above how provisioning can become a cause of *engai* in park-edge settlements. This is likely to lead to affected farmers blaming the park for the damage. In the park-edge theatre crop-raiding monkeys are firmly associated with the parks. Although parks may represent the monkeys as 'wild monkeys', villagers tend to see the monkeys as belonging to the parks that feed them (see Knight 2011: 398–399). Moreover, they readily assume that all monkeys in the area are provisioned monkeys attached to the park. Consequently, any monkey damage around the park tends to be attributed to park monkeys. When local people complain about monkeys, they are implicitly – sometimes explicitly – complaining about the parks deemed responsible for them and their high numbers.

The monkey parks may well dispute this charge. Park managers point out that not all monkey troops in the area are provisioned and that unprovisioned monkeys could therefore be responsible for the damage. After all, most *engai* in Japan exists in localities where there are no monkey parks. In some cases, managers refer back to the existence of *engai* caused by unprovisioned monkeys in the local area in the days before the park was established (and, in part at least, in response to which provisioning was initiated in the first place). But this refusal by the park to accept responsibility for park-edge damage tends to go down badly with affected residents, further polarizing the relationship between the park and the local community.

What this means is that the local perception of *engai* changes. Whereas once *engai* was experienced as a conflict between humans and monkeys, it comes to be seen as a conflict between humans. *Engai* now points beyond the monkeys themselves to the people deemed to have directed or redirected them to the victim's farmland.

Conclusion

In a book on monkey parks in Japan called *Herding Monkeys to Paradise*, I examined how humans control monkey troop movements for the purpose of touristic display (Knight 2011). In this chapter I have returned to the theme of human-induced monkey mobility in Japan by looking at three different responses to the problem of crop-raiding. We have seen that none of these three responses is wholly effective. Blocking monkey access to fields is a hit-and-miss affair because monkeys can learn to penetrate fences or bypass guards. Pushing or chasing away raiding monkeys requires a level of manpower that many Japanese villages today no longer have. And pulling monkeys away from fields to a feeding station, while it may succeed in diverting the animals in the short term, can prove counterproductive longer term as monkey numbers and levels of habituation increase and crop-raiding in park-edge settlements becomes a major problem.

Engai also becomes a conflict between people. Antagonism can arise between farmer and farmer but also between farmers and those involved in monkey tourism. Although in theory the interests of farmers and tourist operators were supposed to become aligned through provisioning, in practice many farmers hold the tourist operators of the park responsible for the monkey damage to their crops. Provisioning was supposed to help tackle crop-raiding by pulling monkeys away from the fields, but it has had the effect of pulling the monkey tourism sector into the crop-raiding problem.

John Knight is Reader in Anthropology in the School of History, Anthropology, Philosophy and Politics at Queen's University Belfast. He is a social anthropologist specializing in rural Japan and his main research topic is human-animal relations. He is the author of *Waiting for Wolves in Japan* (Oxford, 2003) and *Herding Monkeys to Paradise* (Brill, 2011).

Notes

1. By this I mean the key factors affecting monkey movements.
2. Informed consent was obtained from the named informants mentioned in this chapter.
3. On the use of dogs to protect crops from monkeys, see Wada (1998: 98), Yoshida (2012) and Nishiyama (2012).
4. A key consideration here is the effect that provisioning tends to have on reproduction; see Knight (2011: 411–422).
5. Provisioned monkeys in the monkey parks do sometimes act aggressively towards people; see Knight (2011: 233–234, 241–242). However, the level of macaque violence towards people in Japan appears to be considerably less than that reported for some groups of macaques in China (Zhao and Deng 1992).
6. There are, however, other reports that suggest more mixed feelings about monkey culling even in affected communities; see, for example, Watanabe and Ogura (1996: 10).

References

Chūnichi Shinbun. 1997. '"Engai" ni nōka ga himei – Owase' (Farmers' cry for help over monkey damage in Owase), 22 May.

Environment Agency. N.d. *Heisei 25 nendo choju tokei joho* (Statistical information on Wildlife for 2013). Kankyocho. https://www.env.go.jp/nature/choju/docs/docs2/h25/06h25tou.html.

Fukuda, F. 1992. *Hakonesan no saru* (The Monkeys of Hakonesan). Tokyo: Shōbunsha.

Hayashi, K., and K. Watanabe. 2000. 'Chūgoku chihō ni okeru yasei nihonzaru no bunpu dōtai' (The Distribution Trend Among Wild Japanese Monkeys of the Chūgoku Region), in K. Watanabe (ed.), *Honshū no nihonzaru – genjō to hogokanri no mondaiten* (The Japanese Monkeys of Honshū: The Present Situation and Issues of Conservation Management). Inuyama: Kyoto University Primate Research Centre (on behalf of the Working Group on Japanese Monkey Conservation Management), pp. 69–97.

Inoue, M. 2002. *Yama no hatake o saru kara mamoru* (Protecting Mountain Fields from Monkeys). Tokyo: Nōbunkyō.

Itani, J. 1959. 'Engai' (Monkey damage), *Yaen* 4(7): 39–46.

Izawa, K., and MSC (Miyagi-no-saru-chōsakai). 2005. *Saru taisaku kanzen manyuaru* (The Complete Manual on Monkey Countermeasures). Tokyo: Dōbutsusha.

Knight, J. 2003. *Waiting for Wolves in Japan: An Anthropological Study of People-Wildlife Relations*. Oxford: Oxford University Press.

———. 2011. *Herding Monkeys to Paradise: How Macaque Troops are Managed for Tourism in Japan*. Leiden: Brill.

Miller, L.E. 2002. 'An Introduction to Predator Sensitive Foraging', in L.E. Miller (ed.), *Eat or Be Eaten: Predator Sensitive Foraging among Primates*. Cambridge: Cambridge University Press, pp. 1–17.

Ministry of Agricuture. N.d. *Zenkoku no yasei chōjūrui ni yoru nōsakubutsu higai jōkyō (H 19)* (The National Situation of Crop Damage according to Wild Animal Species [2007]). Tokyo: Ministry of Agriculture. Retrieved 25 August 2009 from http://www.maff.go.jp/j/press/seisan/tyozyu/pdf/080919-01.pdf.

Mito, Y., et al. 2000. 'Tōkai-Hokuriku chihō no nihonzaru bunpu henkan' (The Change in Distribution of Japanese Monkeys in the Tōkai-Hokuriku Region), in K. Watanabe (ed.), *Honshū no nihonzaru – genjō to hogokanri no mondaiten* (The Japanese Monkeys of Honshū: The Present Situation and Issues of Conservation Management). Inuyama: Kyoto University Primate Research Centre (on behalf of the Working Group on Japanese Monkey Conservation Management), pp. 42–51.

Muroyama, Y. 2003. *Sato no saru to tsukiau ni wa – yasei dōbutsu no higai kanri* (Spending Time with the Monkeys of the Village: Managing the Damage Caused by Wild Animals). Kyoto: Kyōto Daigaku Gakujutsu Shuppankai.

Nakagawa, I. 2003. *Roman o otte – moto Ōita shichō Ueda Tamotsu monogatari* (Pursuing the Dream: The Story of the Former Mayor of Oita). Ōita: Ōita Gōdō Shinbunsha.

Nishi Nihon Shinbun Chōkan. 1998. 'Utsushiyo no ushirosugata' (The rear view of the picture world), 28 September.

Nishiyama, R. 2012. 'Monkey-chasing dogs in big demand in farm communities', *Asahi Shinbun*, 21 December.

Nomoto, K. 1994. *Kyōsei no fōkuroa – minzoku no kankyō shisō* (The Folklore of Co-existence: The Environmental Thought of Folklore). Tokyo: Seidosha.

Ōita Gōdō Shinbun. 1971. 'Yahari dorubako Takasakiyama' (Takasakiyama – a goldmine), 28 August.

Shikoku Shinbun. 2008. 'Kyūzō – hitozato osou saru no mure' (Sharp increase of monkey troops raiding villages), 22 June.

Sprague, D.S. 2002. 'Monkeys in the Backyard: Encroaching Wildlife and Rural Communities in Japan', in A. Fuentes and L.D. Wolfe (eds), *Primates Face to Face: The Conservation Implications of Human–Nonhuman Primate Interconnections*. Cambridge: Cambridge University Press, pp. 254–272.

Suzaki, S. 2010. 'Engai taisaku ni "nuigurumi" o setchi' (Installing a 'Moving Tiger Doll' as a Monkey Damage Countermeasure), *Chōjūdayori* (Aichi-ken) 2, 21 April.

Takasakiyama. 1998. *Takasakiyama no ayumi* (Takasakiyama's Journey). Ōita: Ōita Takasakiyama Kanri Kōsha.

Wada, K. 1998. *Saru to tsukiau – ezuke to engai* (Interacting with Monkeys: Provisioning and Damage). Nagano: Shinano Mainichi Shinbunsha.

Watanabe, K. 2007. 'Nihonzaru – sansekisuru kadai to kotaigun kontorōru' (Japanese monkeys: Management problems and population control). *Honyūrui Kagaku* 47(1): 147–151.

Watanabe, O., and S. Ogura. 1996. 'Nōson chiiki ni okeru yasei dōbutsu no kachi ninshiki to hogo-kanri seisaku e no ikō' (Relationship between Perceptions toward Wildlife and Opinions about Wildlife Management Policy in Rural Areas of Central Japan), *Wildlife Conservation Japan* 2(1): 1–15.

Yomiuri Shinbun. 1995. 'Nozaru higai, itten "takara no yama"' (From Wild Monkey Damage to a 'Treasure Mountain' in One Go), Ōita edition, 5 February.

Yoshida, H. 2012. *Monkidoggu* (Monkey Dog). Tokyo: Nōbunkyō.

Zhao, Q.-K., and Z.-Y. Deng. 1992. 'Dramatic Consequences of Food Handouts to *Macaca thibethana* at Mount Emei, China', *Folia Primatologica* 58: 24–31.

3
Unintended Consequences in Conservation
How Conflict Mitigation May Raise the Conflict Level—The Case of Wolf Management in Norway

Ketil Skogen

Most conflicts over conservation seem to have obvious 'material cores' in the sense that interests tied to various forms of economically motivated resource utilization or landscape transformation are pitted against conservation. In many cases, these aspects of the conflicts resemble issues that are regularly dealt with within established political and bureaucratic systems, such as those regulating the interaction between the state and the agricultural sector. As a consequence, mitigation efforts normally target such conflict issues, which are comprehensible within a familiar productivist paradigm, and look like they can be addressed by means of conventional measures such as economic compensation, incentives or technical arrangements.

However, many controversies over conservation have additional aspects that are fundamental to conflict development. These are often rooted in other, more general, social tensions and processes of social change that may have less to do with conservation per se. They too certainly have a material basis, but are not necessarily restricted to the effects of conservation or any other specific land management practice. This means that mitigation efforts that primarily address economic loss in a narrow sense (such as lost productivity in agriculture) will often miss these other dimensions. Furthermore, some such efforts may lead to significant unintended consequences, in that they aggravate other aspects of the conflicts.

My example is the conflicts over wolves in Norway. These conflicts are intense, and they are deeply embedded in processes of social change in rural areas. They are shaped by power structures in modern societies, such as those related to class (Skogen and Krange 2003; Krange and Skogen 2007, 2011). Currently, the small Norwegian wolf population is concentrated in areas with limited livestock farming, and some of the most intense conflicts have occurred in places where there is almost no livestock and minimal loss. Here, traditionally oriented factions of the rural working class – not farmers – emerge as a powerhouse of resistance against growing carnivore populations, particularly wolves. This has to do with the problems wolves cause for local hunters, to some extent the fear of wolves that some local people feel, and how the wolf is seen as a symbol of shifting priorities in land management, from resource use to conservation. These changes are not only seen as economically harmful, but as ultimately threatening the whole 'rural way of life', based as it is on utilizing natural resources (Skogen and Krange 2003).

In Norway, conflict mitigation, both in terms of compensation and preventive measures, is almost exclusively geared towards the needs of livestock farmers. As a consequence, these measures, which are officially seen as staple elements in dealing with conflicts over large carnivores, not only alienate people who are supportive of carnivore protection but can also annoy traditionally oriented groups like rural hunters, leaving management agencies and policy makers almost without any allies on the ground. Because these efforts have a narrow, basically economic focus, they produce unintended consequences that are clearly detrimental to conflict mitigation.

Wolf Recovery in Scandinavia

The Scandinavian wolf population is slowly recovering due to protection during the last forty-five years. Sweden and Norway now share a trans-boundary population of approximately 450 animals. About 65 of these live only in Norway and about 25 more have their territories on both sides of the border. There were seven reproductions (litters born) in the Norwegian packs in 2015, and three more in the cross-boundary packs. This is a record high in modern times.

DNA analysis has established that the native Scandinavian wolf population went extinct, most probably in the 1970s. All wolves in the existing population are descended from immigrants from the Finnish-Russian population. Because these have been very few, the current population faces inbreeding problems despite its steady growth (Liberg et al. 2005; Wabakken et al. 2016).

There has been illegal killing of considerable proportions in both countries, as well as legal licenced hunting and culling to protect livestock and Saami reindeer herds. Nevertheless, the wolf population is growing steadily. This means that wolves are reappearing in areas where they have been absent for as long as one hundred years. Wolves have killed livestock, and although the toll has been modest overall, individual herders have been hit hard. They have also killed quite a few hunting dogs, and many more have been wounded. Furthermore, they generate fear among local residents who are not used to having large predators around.

Livestock production is limited in the wolf area, but it does exist. Compensation programs are in place as in the rest of Norway, so farmers suffer no great economic loss even if livestock is killed. Further damage to livestock has been limited partly by extensive and costly preventive measures funded by the government, such as economic support to move sheep, erect electric fences, etc. However, these preventive measures are not directed specifically towards wolves, but rather large carnivores in general.

There is extensive sheep husbandry in other parts of Norway. Here, livestock loss does play a crucial part in the conflicts over large carnivores. The same applies to the reindeer herding areas in the north. But these areas do not currently have wolves. The culprits are lynx, brown bears and, not least, wolverines. These depredation problems are also relatively new, and a result of species protection implemented since the 1960s. However, they started long before the wolves reappeared. Thus, conflict with livestock was already established as a core issue in Norwegian large carnivore management.

Norwegian regional policy is a factor that further strengthens the centrality of the farming perspective in relation to problems with large carnivores. Maintaining rural settlement all over the country has been a stable political goal, and stimulating the agricultural sector and other resource-based industries (forestry, fisheries) has been the cornerstone of this policy (Almås 1989). A high level of subsidies, paralleled only by Switzerland, Iceland, Japan and South Korea (OECD 2012), has maintained active, technologically advanced agriculture spread across much of the country. However, most farms are small by European standards, and in many areas livestock and grass production are the only viable farming activities, due to climatic factors. Because of this, shifting governments have stimulated livestock production, and not least sheep breeding. During the long period when large carnivores were absent (the twentieth century up until circa 1975) husbandry methods entailing free-ranging sheep with limited supervision were developed. Breeding deliberately weakened herding instincts so as to make the sheep disperse across mountains and forests to better utilize the grazing resource and prevent the spread of disease and parasites. Consequently, shepherding and use of guard dogs is now difficult, unless other sheep breeds are used. When large carnivores returned, effects were extremely serious in some areas, and conflicts have flourished ever since.

Due to the active role of the state, a large agriculture bureaucracy exists, and it has its counterpart in large and strong farming organizations. The interaction between the state and the farmers is extremely well regulated, and numerous official communication channels exist. For example, extensive negotiations are held each year between farmers' organizations and the government over subsidies and other issues affecting the agricultural sector. Consequently, there are well-established systems for transferring economic support from the state to the agricultural sector.

Sociological Studies of the Wolf Conflicts

Sociologists from the Norwegian Institute for Nature Research (NINA) have been studying the conflicts over large carnivores within the present distribution range for wolves since 1998. This is an area in southern Norway covering the eastern parts of the counties Hedmark, Akershus and Østfold, all bordering Sweden. Various case studies have been conducted over the years, involving six municipalities scattered along the north-south axis (Skogen 2001; Skogen and Krange 2003; Figari and Skogen 2011; Krange and Skogen 2011).

The northern part (in Hedmark) is sparsely populated. Historically, this has been a forestry region with limited farming. Timber has been the economic backbone, including both the logging itself and forest industry: paper mills, saw mills, wallboard production, etc. As in the rest of the global North, much of this industrial activity has been moved abroad, and the logging is now mechanized, employing only a tiny percentage of the workforce that was needed as late as in the 1960s. To some extent this has been compensated by growth in both the public and private service sectors, including tourism, but – except for a few semi-urban centres – this region has seen continuous depopulation for decades.

If we move south through the wolf region, the picture gradually changes. In Akershus, the municipalities that are now, or have at one time been, home to wolves, are also heavily forested, although the presence of farming on a larger scale increases as one moves south into Østfold. Østfold, in the extreme southeastern corner of the country, also has forests. Along the Swedish border, these are relatively continuous like further north, but interestingly some of the wolves seem to have opted for forest patches scattered within an agricultural landscape. The presence of agriculture is much stronger than in the north, both physically and economically, but there is limited livestock production and no rough grazing in this grain-producing area. In Akershus and Østfold, many people live within comfortable commuting distance to urban centres, including Oslo.

Our studies have clearly demonstrated that conflicts over wolves are *social* conflicts. We have seen that people who oppose wolf protection are often much angrier with their human adversaries than with the animals, and the conflicts reach far beyond controversies over management practices: they are about wider processes of social change perceived as threatening by many people in rural areas. Anti-wolf attitudes predominantly prevail among people who are firmly rooted in traditional land use practices and in a rural working-class culture. These attitudes are not always – or even predominantly – related to adverse material effects of wolf presence experienced at the individual level. Granted, wolves are seen as a threat to rural lifestyles because they kill hunting dogs, decimate local moose and roe deer stocks, and cause some fear among locals. But few people have ever encountered wolves or had their dog attacked, and the effects wolves have on game populations vary a good deal (Gundersen et al. 2008; Gervasi et al. 2012).

Evidently there are aspects of these conflicts that are not directly related to the wolves. We think general development trends in rural areas constitute a key factor. Government efforts to maintain rural settlement are not as effective (or as extensive) as they once were; even rural Norway is exposed to the forces of economic modernization and globalization. Although still better off than rural areas in many other countries, depopulation and economic decline are increasingly prominent features. The dominant narrative among people with cultural ties to the resource-based economy is one of economic decline, leading to depopulation and dismantling of

Unintended Consequences in Conservation 53

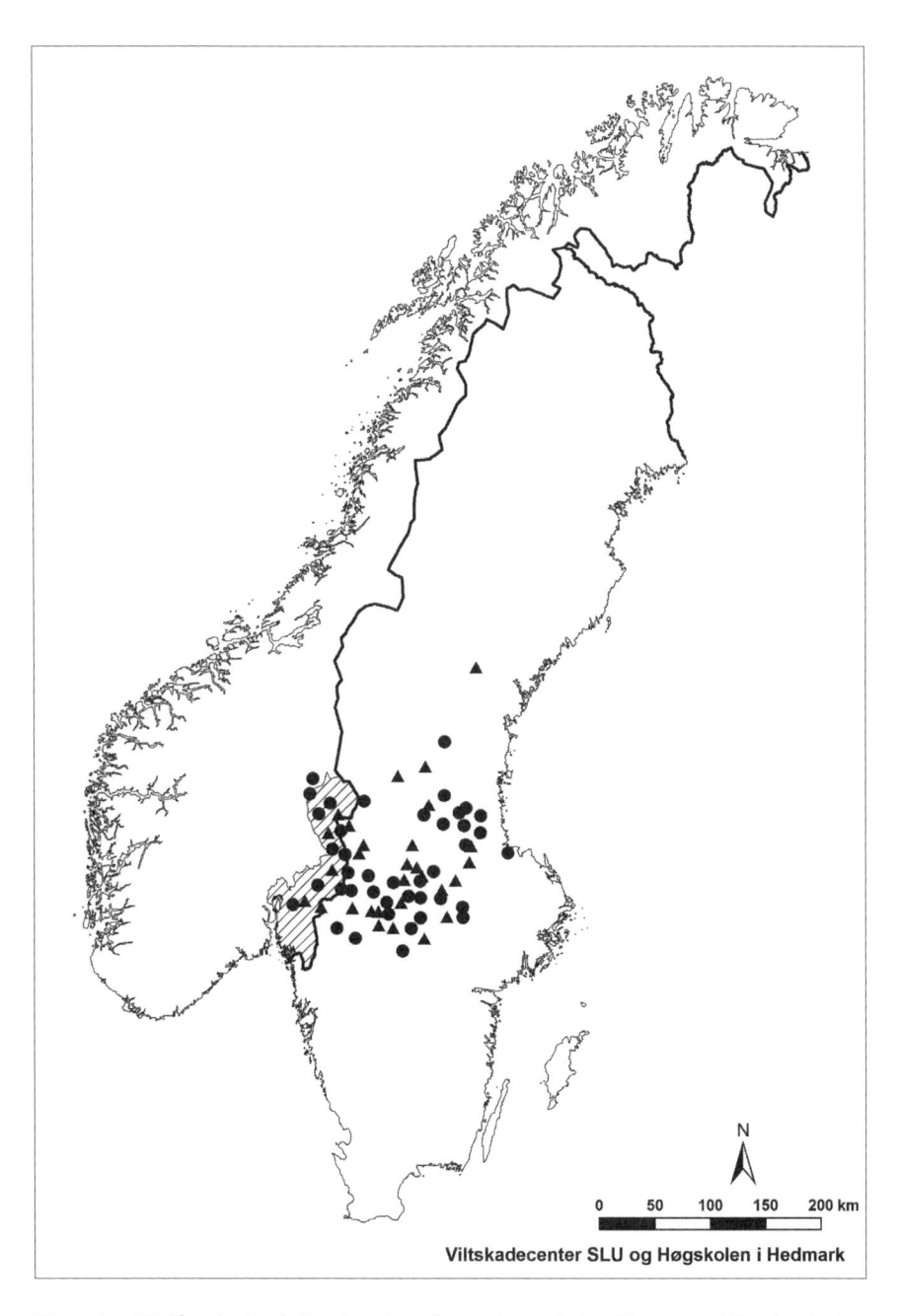

Figure 3.1. Wolf packs (circles) and territorial pairs (triangles) in Norway and Sweden during winter 2015–2016. The Norwegian management zone for wolves is indicated (hatched). *Source*: Rovdata – the database for the Norwegian Large Predator Monitoring Program (http://www.rovdata.no/Ulv/Bestandsstatus.aspx).

private and public services (Skogen, Mauz and Krange 2008; Krange and Skogen 2011). The forest industry employs only a handful of people, and the small-scale agriculture, which was typical of marginal areas, is disappearing fast. Farm abandonment leads to spontaneous reforestation of fields that were highly valued by local people because they opened up the landscape (see also Ghosal, Krishnan and Skogen 2015). Importantly, this happens in a time when a conservation ethos has achieved a hegemonic position in public discourse, and increasingly manifests itself in practical land management: restrictions on land use (e.g. logging) are implemented, new protected areas are established and species that were previously persecuted are now protected. Some social groups interpret these changes in the cultural valuation of nature (of which wolf protection is one striking expression) as a driving force behind the decline in resource industries, and as a threat to a traditional rural lifestyle that rests on a resource economy and entails forms of outdoor recreation based on harvesting (Krange and Skogen 2011).

But the picture is not clear-cut, as the population is diverse even in these areas, and pro-wolf attitudes are certainly present. Most rural communities are now more economically diversified than before, with considerable proportions of the population employed in the service sector (including tourism) and in the public sector. Middle-class jobs demanding higher education have increased in number in public and private sector service provision, significantly changing the educational profile of the work force. Accordingly, and also due to increased and often extensive contact and exchange with urban areas, and the evolution of new media, cultural diversification has occurred. For a considerable proportion of the population, even in rural areas, the utilization of nature is relegated to leisure. Traditional consumptive recreation activities may lose their significance, being bound to livelihoods entailing physical interaction with nature. Even in small rural communities we now find landscape interpretations that embrace wilderness, as opposed to the production landscape. From such a perspective, human activities (resource extraction tasks as they are performed today) are seen as harmful. This way of seeing the landscape and resource-related tasks is predominantly found among people who are not culturally rooted in traditional land use. To them, the wolf can be a strong symbol of an authentic, wild nature that preceded the human-dominated (and now partly damaged) landscape (Figari and Skogen 2011; Ghosal, Krishnan and Skogen 2015). Their interaction with the land entails valuing it through non-consumptive recreation and as a symbol of the unspoiled, something that should be revered and left in peace. They thus deviate from what one may term a traditional landscape interpretation, where human appropriation of nature is seen as necessary and benevolent – not only to people, but also to wildlife and the land itself (Krange and Skogen 2011; Skuland and Skogen 2014).

Just as the concept of wilderness is tied to the idea of what once used to be, the notion of productive nature is associated with continuity and heritage from earlier generations. For people rooted in traditional, resource-based land use, the traces of ancestors' hard work and efforts to *tame* the wilderness express the inherent meaning of the physical environment, and must be preserved through continuation of traditional practices. This cultural landscape must be saved from 're-wilding'. Domestic animals as well as huntable game must be protected against predators. Seen from such a perspective, humans and nature are not separate: traces of human activities are no more negative than traces of other beings that belong on the land (Figari and Skogen 2011; Ghosal, Krishnan and Skogen 2015). But it also means that humans can legitimately control creatures that cause problems and that do not perform any meaningful function in this production landscape.[1]

Controversial Conflict Mitigation

Huge resources are used on compensation and preventive measures directed at the livestock sector, although it is a limited economic activity in the current wolf area. I shall now describe the most important measures that have been taken, and explain why they have been controversial.

Zoning and Regional Management

Norway is presently divided into eight management regions for large carnivores. Some authority over management has been transferred to regional boards that are politically appointed (by the county assemblies) in an effort to bring decision-making closer to the people who are affected, and to introduce an element of 'local democracy' to carnivore management. The ultimate goal is conflict reduction through increased legitimacy. The system, which was implemented in 2005, has not been properly evaluated, but research indicates that many local people do not see the boards as being either sufficiently local or under any meaningful democratic control (Skogen, Krange and Figari 2013). These regional bodies (with technical and scientific support from the county environmental agencies) are now responsible for managing brown bears, lynx and wolverines, within a nationally established framework entailing, for example, population goals for each region and species. However, a special management zone for wolves has been established outside this system, due to particular management challenges – including the virtual impossibility of combining territorial wolf packs and free-ranging sheep. This zone partially covers two ordinary zones, and involves four counties. It is managed by a special board, made up of people from the two involved regional boards. As one can see, this is a complicated system, not likely to alleviate people's sense of alienation towards management institutions (Skogen, Krange and Figari 2013). Furthermore, the boards cannot go outside their specific mandate; if they do, the national Environment Agency will take over. Their decisions on hunting quotas that they may set for lynx, bears and wolverines, as long as population goals have been met, can be appealed by conservation groups and overturned by the Ministry of the Environment (which has happened on several occasions). So even though the boards are often regarded as predominantly 'anti-carnivore' (no stakeholder representation here, only political appointments), their establishment has not changed the management regime on the ground in any major way (which a number of local people have observed, and which causes frustration for those who would like to see fewer large carnivores around) (Skogen, Krange and Figari 2017).

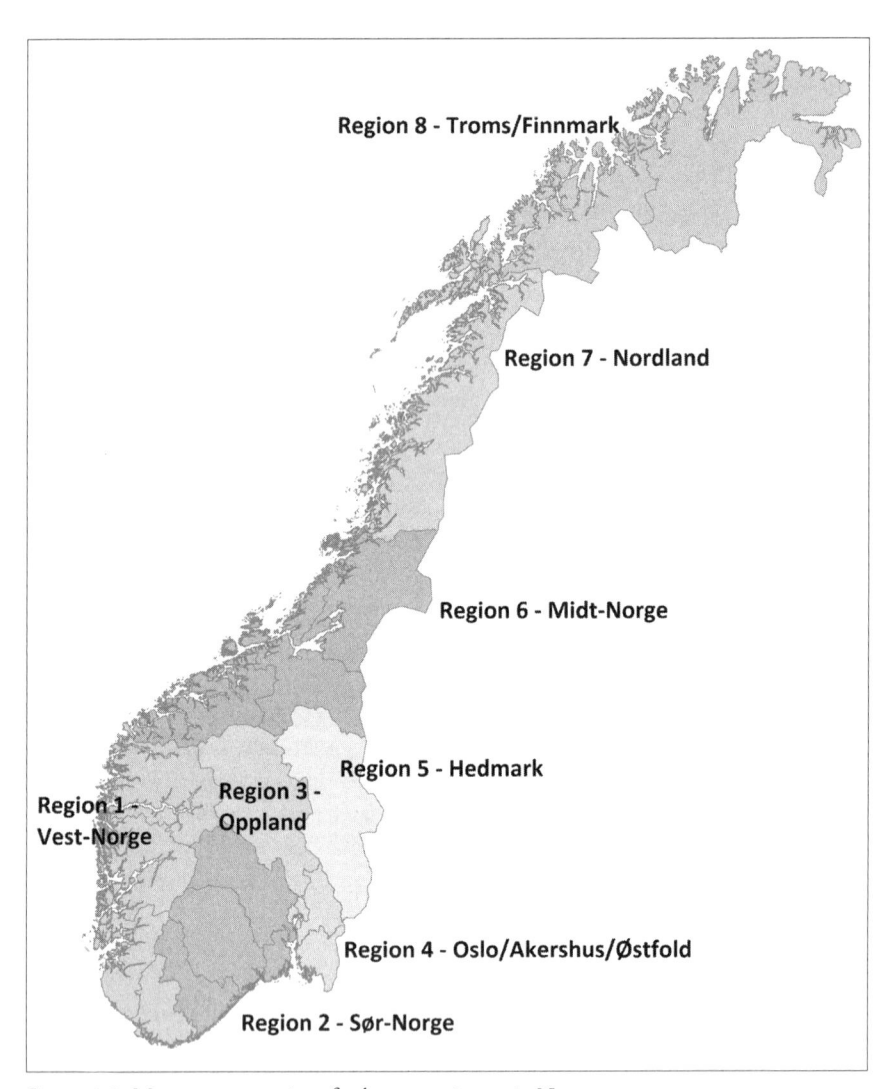

Figure 3.2. Management regions for large carnivores in Norway.

For the wolf-management zone, the general principle is that wolves are allowed to establish territories and breed inside it, and the threshold for culling problem animals will be high. Outside the zone, however, this threshold is lower, and repeated attacks on livestock will trigger culling. There is also some limited licensed hunting outside the zone; the quota was six animals in the 2015/2016 season (all shot during the first few weeks).

The management zone for wolves has been delineated specifically to minimize livestock losses. However, this has only been possible because the wolves are already concentrated in the border areas, close to Sweden. When the new immigrants first crossed the border, they arrived in forest areas where farming has never been economically important, and in agricultural areas with limited livestock husbandry. There are some enclaves in the northern part of the zone where sheep farming exists, but compared to western and northern Norway, it is still very limited, involving few farmers and relatively small flocks.

The wolf zone is widely ridiculed by residents within it, and by many who live outside. 'Do they think the wolves can see the boundaries?' is a common expression, regardless of whether people are for or against wolves. People understand the zoning logic as an attempt to contain wolves in a specific area, which is seen as ludicrous because wild animals will go wherever they want to. Authorities deny this is the purpose of the wolf zone. According to them the zone is mainly an area where the wolves have stronger protection than outside. No efforts are made to prevent wolf movement, but a different management regime applies outside the zone. However, this more nuanced explanation does not seem to impress people who see the zone as both stupid and – importantly – unfair (Figari and Skogen 2008, 2011; Skogen, Krange and Figari 2017).

Those who are positive towards wolf presence tend to see it from the wolves' point of view. The zone is too small for them, each pack needs a large territory and wolves should be able to wander wherever it is natural for them to go. So the zoning is an example of anthropocentric priorities that favour human needs, at the expense of wildlife and nature in general. Seen from a different perspective, namely that of the wolf opponents, the zone is also too small, but that is because the burden of having wolves is unevenly distributed. If we are to have wolves in Norway at all, the problems should not be heaped on the few farmers and the many hunters who live inside the zone.

So we may conclude that everybody who cares one way or the other and who lives inside the zone is frustrated. The zone is also widely ridiculed by people who live outside it, but some farmers and hunters see it as a second-best solution as long as they do not have to cope with wolves on a regular basis. Certainly, this is not a typical example of a successful conflict mitigation strategy, especially since biologists also see the zone as far too small and ridiculous from an ecological perspective.

Fences

To minimize attacks by carnivores on sheep, different fencing methods have been implemented. There are several varieties of electric fences meant to keep predators away from the sheep. Some of these have been erected around cultivated fields, entailing a grazing strategy that differs from the one that has dominated for decades. Farmers see a number of problems with keeping sheep in the fields, including the spreading of disease and parasites. According to the farmers we interviewed, this is a big problem in parts of the world where this is the normal way to graze sheep. The worst effect would be the loss of the vast backcountry grazing resource. There have also been many technical problems with the fences themselves, and they have not generally been effective in keeping predators out. First-generation fences had design faults, and some fences were not built correctly or were poorly maintained.

Even though substantial government support has been provided for the erection of fences, maintaining them and vaccinating sheep means extra work for the farmers. Consequently, this type of fence has not been popular with sheep owners, even though some have given up their resistance and let their sheep graze in fields near the farm (a practice that in itself is entitled to economic support because it is a preventive measure).

To overcome some of these problems a different approach has been tried out in two places within the wolf zone. So far, two large forest areas have been fenced in Trysil (a NINA study area) and Grue, both in Hedmark, so that grazing inside the fences would mimic the normal rough grazing. These fences were erected following a long and turbulent process, particularly in Trysil: many landowners were sceptical for various reasons, and rejected fences on their property even though some sheep owners had historical grazing rights there. This ruled out the inclusion of some of the more productive grazing areas that have traditionally been used. There was also disagreement among sheep owners, as some thought moving their sheep to the fenced area would be too time-consuming, and claimed that the grazing there was inferior. It was also claimed that accepting fences would amount to defeat in the bigger battle against having carnivores in the region. Eventually the fence in Grue was finished in 2007 and in Trysil in 2008, paid for in full with government funding for conflict mitigation.

These fences are formidable structures: 130 cm tall, with six electric wires at a voltage of 4,500 V. The enclosed area in Trysil is 2,200 ha and in Grue 1,100 ha. For maintenance purposes, roads are needed all along the fences, amounting to 23 km in the Trysil case. Neither of the fences was effective in their first couple of years. In both cases, large carnivores, bears and wolves, got inside and caused more livestock deaths than they would have managed outside the fence. This caused enormous controversy and media buzz, particularly in local newspapers and regional broadcasting, exacerbated by the fact that one bear was a reproductive female that wildlife managers decided to move by helicopter (instead of culling it). This was obviously a very costly undertaking. However, after having, apparently, solved some technical problems – e.g. related to voltage drop – both fences proved effective in the 2011 grazing season.

So the fences were not successful for the first few years from a sheep-herding viewpoint. Many sheep and predators were killed. As I have stressed several times already, the wolf zone is not really a sheep-producing region, and relatively few farmers have been involved in the fencing projects. What we observe here is that substantial government funding is directed towards alleviating the problems of a relatively small group. Granted, it is a group whose problems are potentially of considerable impact to their livelihoods, so this should be a good thing surely? However, the fences have been met with considerable resistance locally (Skogen, Krange and Figari 2017).[2]

These fences were seen by many local residents as destroying pristine landscapes, primarily because of the additional road construction associated with them. This was perhaps most evident in the Trysil case, where the area that was eventually selected (the only area that could be obtained from landowners) was seen by many as one of the few relatively intact wilderness areas within the municipality's borders. This view was evident from news coverage and letters to editors, and was also demonstrated by research in the area prior to the erection of the fence (Skogen, Krange and Figari 2017). In addition to the aesthetic impact on the landscape, there was great concern about the effects the fences would have on wildlife movement. Particular concern was voiced concerning moose calves that were born inside the fence and would be unable to get out. The company that makes the fences claimed that adult moose and roe deer would have no problems jumping the fence, but this claim was also doubted, particularly by some of the hunters we interviewed. (However, experience from the last grazing seasons seems to have proven the constructors right – and there have been no reports of trapped calves either.) Some of the hunters also mentioned that the most vulnerable species would be the sheep themselves. According to these hunters, sheep are widely seen as stupid and are known to get tangled in ordinary fences. So what to expect when they had to deal with a high-voltage fence? And, importantly, many interviewees, across focus groups, were worried that the fences might impair the free access that Norwegians are accustomed to. There is a general right of access even on private land, according to the 'right to roam' legislation, which is based on customary rights that were never challenged in Norway. This would not least be harmful to hunting, according to many locals who were not involved in farming themselves. A limited number of access 'bridges' were established but many would disagree that this amounts to free access. The fence in Trysil is designed to be dismantled in the winter, but the wires and poles will remain on the ground, allegedly posing danger to wild animals. Accordingly, the powerful Trysil chapter of the Norwegian Association of Hunters and Anglers issued several sharp statements, in the form of letters to various authorities, media statements, etc., and put considerable pressure on the municipal council to stop the fence.

In our research material from Trysil, the most vocal opponents to the fence were hunters. A general impression from our interviews with hunters was that this issue evoked strong feelings – quite a bit of aggression surfaced when fencing was discussed. There were certainly also doubts among non-hunting outdoors enthusiasts and people with conservationist leanings, but a more positive attitude seemed to prevail among these: anything to reconcile adversaries in the heated conflict. However, many hunters, and their organization, were extremely outspoken. This is interesting, given that many hunters are at least as strongly opposed to current carnivore protection as are farmers, and hunters' organizations tend to support this view. Indeed, the arrival of large carnivores, and particularly wolves, has forged an alliance between livestock farmers, landowners and rural hunters (who often have a working-class, non-farming background) that was not there before (Skogen and Krange 2003). This alliance bridges (or covers up, depending on one's perspective) conflicts that have much deeper roots than the current appearance of harmony. These have historically been related to land ownership versus hunting and grazing rights, friction between hunting and sheep herding (sometimes involving dogs), and general class relations in rural areas (Skogen and Krange 2003).

The fence issue seems to be driving a wedge in the – probably quite fragile – alliance between hunters and farmers on the carnivore issue (see also Skogen and Krange 2003); but not because the hunters are in favour of the wolves all of a sudden (not all hunters are anti-wolf, but I do not go into that here). Granted, they draw the same conclusions pertaining to the fence itself as do the vocal pro-wolf groups. (The latter were more active in Grue, where they ran an anti-fence campaign on a pro-wolf website, and were at one point accused of sabotaging the fence, according to interviews with farmers in the local newspaper Østlendingen). The same concerns were voiced in a more subdued way by outdoors enthusiasts and moderate conservationists in Trysil. They observe exactly the same negative effects for wildlife, humans and landscape, and emphasize those in similar ways. Yet they come to very different conclusions concerning the wolves' role in all this: from an anti-wolf perspective, the need for fences simply emphasizes that the wolves do not belong here anymore. But everybody is angry at the fences, and at the agencies that have promoted and funded their installation.

Monitoring and Hands-on Management

The detailed goals that are set politically for population size and distribution of large carnivores clearly demand precise knowledge that can only be obtained by extensive monitoring. This is achieved by means of several methods, most prominently snow-tracking, DNA-testing and GPS-collaring. But telemetry and other 'hands-on' management practices provoke wolf lovers and wolf adversaries alike – principally for the same reasons. While developed for research purposes, these methods are increasingly used for monitoring that is motivated exclusively by management needs. Again, these needs are shaped by the emphasis on the problems experienced by the farming sector. This has necessitated a form of micro-management where minute knowledge of populations' development and distribution – almost down to each single individual – is essential. But the extensive use of hands-on methods, e.g. the use of GPS collars, which entails chasing (often by aircraft), sedating and capturing animals, is seen as problematic by many people who have strong opinions on large carnivore management. Our studies have shown that both supporters of large carnivores and their opponents tend to agree that animal welfare problems arise, and – importantly – that the extensive use of technology and hands-on procedures transform animals from wild to non-wild (Figari and Skogen 2011; Skogen, Krange and Figari 2017).

Interestingly, wolf supporters and wolf sceptics speak about the wolf itself – not to be confused with its presence in Norway today – in ways that are very similar to each other (Figari and Skogen 2011). Nobody sees themselves as hating the animal: wolves in their natural environment are seen as impressive and fascinating, they are intelligent, social – and above all – *wild*. The core of this disagreement is whether wolves belong in Norway today; in a landscape that may look like a wilderness to some, but is actually a managed production landscape (for timber, grazing, and other uses).[3]

Despite this important discrepancy, there is a relative consensus that intensive monitoring threatens the very notion of a wild animal, at least when this is a routine procedure, conducted for management purposes, and not for research. However, this agreement does not mean that everybody looks upon wolf presence in the same way – as I have already explained. In parallel to the fence issue, those who think there are already too many wolves will tend to see the need for extensive and invasive monitoring as proof that Norway is not a wild place, and therefore has no room for wolves. Those who see the wolves in a positive light, see the monitoring as destructive to a wildness they would like to see in Norwegian nature.

Furthermore, many see the extensive monitoring as a tremendous waste of money. The same argument was raised concerning the fences, but not as vocally as in connection with monitoring. The resources spent are seen as out of all proportion in relation to other needs in Norwegian society, including research and management related to other wildlife species. Indeed, huge amounts of government money are being spent on large carnivore management in Norway. The bulk of this is directed towards compensation and preventive measures, for example moving sheep and constructing fences, but monitoring is being assigned increasing amounts of funding. Additionally, research funded by management agencies is extensive. This is because the conflict level is regarded as unacceptably high, and – importantly – because one party to these conflicts is the farming sector, with its strong organizations, matching state bureaucracy and its longstanding political influence.

Interestingly, what we see again is that the opinion that large carnivore management is too expensive is spread across all positions in the controversy. The reasons are those we discussed earlier. If you want more wolves and savour the wildness they strongly symbolize, you do not want them to be controlled like livestock or pets. They should be their own masters, truly wild animals in a wild nature. This means that large sums are wasted on an essentially immoral activity. And to those that would like to see smaller populations than today, then – again – the need for extensive monitoring is proof that these animals do not belong, or at least that their numbers should be brought down. Achieving that should be easy, according to many interviewees, because large carnivores were almost exterminated a century ago. This could be repeated without spending huge sums that could be used on grouse research or caring for the elderly.

Unintended Consequences

Employing an economic rationale when attempting to mitigate conflicts can appear to be the obvious strategy: somebody's material interests are threatened, so we try to prevent loss. If we cannot, then we compensate the loss. But what if the conflicts are really about something else for a lot of people? What we have seen in Norway is that no matter how much money is poured into certain forms of mitigation, significant conflict dimensions either stay the same or escalate.

Conflict mitigation, collaborative management, etc., in resource management often seems to be aimed at making everybody happy, or at least to distribute frustration equally. But this is rarely possible, and our case in point demonstrates how established power structures (tied to government-agriculture interaction) influence the outcome. It is no coincidence that one small group receives so much of the attention (and indeed of the money).

The unintended consequences materialize when other groups are aggrieved by mitigation efforts directed at a particular sector; the livestock farming sector in our case. If these 'other groups' are those advocating for strong wolf protection, then frustration on their part must be expected as part of a compromise. Conservationists are certainly not happy with current management of large carnivores in Norway, and as we have seen, some conflict mitigation efforts have triggered reactions from that camp. But in addition, many people who do not have favourable views on large carnivores in the first place are also alienated by core elements in the current conflict mitigation strategy. Although this may lead to friction between groups that have been allies in the battle against wolf protection, such as hunters, landowners and sheep farmers (Skogen and Krange 2003), it does not mean that the non-farmers are driven into the arms of the conservationists. They do not change their views on wolf management in Norway, but they become even more frustrated with the management system, the policy makers and the social groups they associate with the increasingly powerful conservation paradigm (i.e. middle-class academics with an urban background) (Skogen and Krange 2003; Krange and Skogen 2011).

Large carnivore policy and management suffers from a lack of legitimacy in diverse social groups, some of which have quite different views on the subject. If norms or laws lack legitimacy, people do not feel an obligation to obey them. It may be that large carnivore management and the legislation that supports it are developing into an area where quite a few people think the government is devoid of legitimate authority. This obviously hampers dialogue, which is problematic in itself. Even graver consequences may follow if a perceived lack of legitimacy is used to defend criminal acts such as illegal hunting – or, for that matter, intentional disruption of legal hunting (which has also occurred in both Norway and Sweden). This in turn inevitably leads to sharper antagonism, for example between conservationists and rural hunters. And that is indeed a tragic development, as these groups in many ways have a common interest in conservation, and should ideally be fighting together against habitat destruction and biodiversity loss.

In this chapter I do not suggest any specific solutions to this dilemma. My intention is to highlight a phenomenon that is probably quite common in conservation controversies. The main message is that we need to look carefully at the wider social context of these conflicts, and get rid of the illusion that they can be 'resolved' solely through the use of tools that are at the disposal of management agencies and policy makers in this particular field. Conflicts – or rather aspects of them – can surely be alleviated, but not in such a way that all groups are equally pleased or displeased by the measures that are adopted.

In the Norwegian wolf example, it seems that the farmers, who get most of the attention now, are simply not the most important group to reach out to in order to decrease the level of conflict specifically related to the wolves, and/or to build sustainable alliances for the greater conservation good. Other social groups play more significant roles in the wolf controversy, and might possibly have a bigger potential as conservation allies – but that is another matter, outside the scope of this chapter.

Ketil Skogen is a sociologist who has done extensive research on conservation conflicts, particularly those tied to the conservation of large carnivores, in a context of social change, power relations and class cultural friction.

Notes

1. The Norwegian word *utmark* embodies this notion of the *used* land, but at the same time land that is cared for. This is land that is not cultivated, but still used (or has the potential for being used). The word *villmark* (which translates directly into 'wilderness') means something else entirely. In our interviews, many rural residents used the word *utmark* with passion, and felt that the word *villmark* was inappropriate, even insulting.
2. Further north, in Trøndelag, outside the wolf zone, a different fencing method has been tried (meant to keep lynx and bears out). This entails fencing areas that are smaller, and which comprise both fields and forest within the same enclosure. These areas are close to the farms, so they are easily accessible to the farmers, and local people are used to fences there: fences near farms are normal. Because they are shorter, these fences are also easier to maintain, and thus they are more efficient. These fences appear to have been more effective, and have been well received by local non-farmers. However, there are concerns about over-grazing in the long term within the relatively small enclosures.
3. See Ghosal, Skogen and Krishnan (2015) for discussion of similar social constructions of the landscape in India and Norway, and consequences for attitudes towards large carnivore presence.

References

Almås, R. 1989. 'Characteristics and Conflicts in Norwegian Agriculture', *Agriculture and Human Values* 6(1): 127–136.

Figari, H., and K. Skogen. 2008. *Konsensus i Konflikt: Sosiale Representasjoner av Ulv*. Oslo: Norsk Institutt for Naturforskning.

———. 2011. 'Social Representations of the Wolf,' *Acta Sociologica* 54(4): 317–332.

Gervasi, V., et al. 2012. 'Predicting the Potential Demographic Impact of Predators on Their Prey: A Comparative Analysis of Two Carnivore-Ungulate Systems in Scandinavia,' *Journal of Animal Ecology* 81(2): 443–454.

Ghosal, S., K. Skogen and S. Krishnan. 2015. 'Locating Human-Wildlife Interactions: Landscape Constructions and Responses to Large Carnivore Conservation in India and Norway', *Conservation and Society* 13(3): 265–274.

Gundersen, H., et al. 2008. 'Three Approaches to Estimate Wolf Canis lupus Predation Rates on Moose Alces alces Populations', *European Journal of Wildlife Research* 54(2): 335–346.

Krange, O., and K. Skogen. 2007. 'Reflexive Tradition: Young Working-Class Hunters between Wolves and Modernity', *Young – Nordic Journal of Youth Research* 15(3): 215–233.

———. 2011. 'When the Lads go Hunting: The "Hammertown Mechanism" and the Conflict over Wolves in Norway', *Ethnography* 12(4): 466–489.

Liberg, O., et al. 2005. 'Severe Inbreeding Depression in a Wild Wolf Canis lupus Population', *Biology Letters* 1(1): 17–20.

OECD. 2012. *Agricultural Policy Monitoring and Evaluation 2012*. Paris: OECD Publishing.

Skogen, K. 2001. 'Who's Afraid of the Big, Bad Wolf? Young People's Responses to the Conflicts over Large Carnivores in Eastern Norway', *Rural Sociology* 66(2): 203–226.

Skogen, K., and O. Krange. 2003. 'A Wolf at the Gate: The Anti-Carnivore Alliance and the Symbolic Construction of Community', *Sociologia Ruralis* 43(3): 309–325.

Skogen, K., I. Mauz and O. Krange. 2008. 'Cry Wolf! Narratives of Wolf Recovery in France and Norway', *Rural Sociology* 73(1): 105–133.

Skogen, K., O. Krange and H. Figari. 2017. *Wolf Conflicts. A Sociological Study*. New York: Berghahn Books.

Skuland, S.E., and K. Skogen. 2014. 'Rovdyr i Menneskenes Landskap', *Tidsskriftet Utmark*. http://utmark.nina.no/Portals/utmark/utmark_old/utgivelser/pub/2014-1%262%26S/ordin/Skuland_Skogen_UTMARK_1%262.html

Wabakken, P., et al. 2016. *Ulv i Skandinavia og Finland: Sluttrapport for Bestandsovervåking av Ulv Vinteren 2015–2016*. Elverum: Høgskolen i Hedmark.

4
Badger-Human Conflict
An Overlooked Historical Context for Bovine TB Debates in the UK

Angela Cassidy

Since the early 1970s, the question of whether to cull wild badgers (*Meles meles*) in order to control the spread of bovine tuberculosis (bTB) infection in British cattle herds has been the source of public controversy. Bovine TB is caused by *Mycobacterium bovis*, a microorganism that can in principle infect any mammalian species including humans, although its main host is the domestic cow. In the United Kingdom *M. bovis* was a major cause of tuberculosis in humans until well into the twentieth century, as it can be transmitted zoonotically via infected meat and milk in particular. The gradual recognition of this link by scientists, veterinarians and public health authorities led to the establishment of many modern systems for regulating food risks, including the pasteurization of milk, meat inspection and routine TB testing of cattle herds (Atkins 2000; Waddington 2006). Due to the success of these systems, in many countries bTB no longer poses a serious public health threat; however, on a global scale it still contributes to human disease, particularly in several African countries, and in specific populations worldwide (Müller et al. 2013). Despite this, bTB is still a major economic problem in the United Kingdom, as cattle testing positive for the infection are slaughtered and farmers must be compensated by government for such losses. Furthermore, herds containing infected animals are placed under movement restrictions, meaning farmers also suffer significant disruption to their livelihoods, and associated stress and emotional fallout from this disruption and the loss of their animals (Mort et al. 2008; Farm Crisis Network 2009). Despite the near eradication of the disease by the late 1960s, bTB infection rates in British cattle slowly crept back up through the second half of the twentieth century, accelerating steeply following the 2001 outbreak of foot and mouth disease (DEFRA 2014).

While the full reasons for this resurgence remain unclear, veterinarians and farmers have pointed to the existence of a 'reservoir' of infection in wild badger populations, and have lobbied for bTB management policies to include badger culling to remove this source. At the same time, conservation and animal welfare groups have contested the importance of wildlife reservoirs, pointing instead to farming practices as a potential cause, and have campaigned against culling policies. Following a review of the scientific evidence (Krebs 1997), the UK government commissioned the Randomised Badger Culling Trial (RBCT), designed to test the effects of badger culling on bTB rates in cattle through a systematic field trial. After nearly ten years of intensive research carried out over about 100 km² of the British countryside at a cost of about £50 million, the multidisciplinary research team conducting the study concluded that some forms of culling appeared to facilitate the spread of bTB, and that 'badger culling cannot meaningfully contribute to the future control of cattle TB in Britain' (ISG 2007: 14). However, the RBCT findings failed to resolve the controversy, and instead these conclusions were publicly contested by veterinary and farming associations, as well as the government's own Chief Scientific Adviser at the time (see Cassidy 2015 for further analysis). The long-term effects of the RBCT culls, alongside their significance and implications for both disease ecology and bTB policy, continue to be hotly debated in the scientific literature (e.g. McDonald 2014; Boyd 2015; Donelly and Woodroffe 2015). In policy and the wider public sphere, we now see a situation where advocates both for and against badger culling argue that their positions are supported by 'sound science'; as have the full spectrum of policies implemented across the United Kingdom since 2008 (Spencer 2011; Lodge and Matus 2014; Robinson 2015).

This chapter presents findings arising from a broader research programme investigating the history of bTB in the United Kingdom since the 1960s as a case study of public scientific controversy – where controversies between scientists take place in the wider public sphere (e.g. Cassidy 2006). It takes a step back from questions of animal health policy to focus instead on the wild animals at the centre of this debate. Bovine TB is a global disease problem, and several countries have both reservoirs of infection in wildlife and active culling policies without attracting the degree of public opposition experienced in the United Kingdom (More 2009; Hardstaff et al. 2014). The European badger (*Meles meles*) is a member of the mustelid family (alongside weasels and otters), although historically it was thought to be a bear (see figure 4.1; also Pease 1898). Badgers are omnivorous, nocturnal foragers that in Britain live underground in large family groups; these groups defend well-defined, stable territories. The range of *Meles meles* extends all the way from Spain in the west to Iran in the east; and south to north from Spain to the Arctic Circle (Roper 2010: 12).

Despite the coexistence of both badgers and bTB across much of this area, it is only in the United Kingdom and the Republic of Ireland that direct causal links have been drawn between these wild animals and infections in domestic cattle. This suggests that a specific combination of ecological, economic, social and cultural factors contribute to the disease ecology of bTB in these countries (Woodroffe et al. 2009; Roper 2010: chap. 7; Byrne et al. 2012; O'Connor, Haydon and Kao 2012; Atkins and Robinson 2013a, 2013b; Fitzgerald and Kaneene 2013). Across much of Europe, population densities are low and in some countries badgers are hunted

Figure 4.1. Anti-cull graffiti, Leamington Spa, United Kingdom, July 2015. *Source*: author's own photograph

animals; however, in Britain and Ireland populations are thought to be much higher. The United Kingdom and Netherlands are the only two countries where badgers benefit from specific legal protection, alongside a network of local groups concerned for their welfare and conservation (Griffiths and Thomas 1997; Runhaar, Runhaar and Vink 2015). Despite this, in the United Kingdom these animals continue to be

68 Understanding Conflicts about Wildlife

subject to (illegal) human practices of 'badger baiting' (fighting for sport), digging (digging out a sett and/or sending dogs in to hunt the animals) and 'control' activities from farmers and gamekeepers (Roper 2010: 39–41; Enticott 2011a). Alongside the similarly conflicted fox (Woods 2000; Marvin 2001) and otter (Allen 2010; Syse 2013), badgers are highly culturally significant in Britain, appearing in popular folklore, fiction, poetry and visual imagery throughout the twentieth century.

This chapter provides an analysis of these cultural sources, alongside contemporary media coverage of debates around badger culling to ask why proposals to cull this wildlife species have provoked such intense and sustained controversy in the United Kingdom of the late twentieth and early twenty-first centuries. Existing social science investigations of the UK bTB problem have focused upon animal health policy and governance (Wilkinson 2007; Grant 2009; Spencer 2011; Maye et al. 2014), the roles of scientific knowledge (Enticott 2001), farming (Enticott 2008), veterinary (Enticott 2011b) and public (Enticott 2015) perspectives on the issue. A few studies have addressed the wildlife aspects of the debate and/or how the issue has played out in the broader public sphere of media and popular culture (Lodge and Matus 2014; Naylor et al. 2015). In common with many environmental controversies, mass media have provided a central arena for debating badgers and bTB, where the various actors involved in the controversy have engaged with wider publics alongside policy makers and politicians (Lester 2010: 37–58). As Molloy (2011) and Corbett (2006: chap. 7) point out, animals often play important roles in developing engaging and appealing mass media content; in turn these media representations contribute to broader public discourses and understandings of animals' roles in human societies. This work has also drawn upon the developing field of animal studies, where researchers working across a range of social science and humanities disciplines have investigated many aspects of human-animal relations, from pet-keeping, farming and sport through to eating (see, e.g. Buller 2013, 2015). The animal studies literature has seen a particular emphasis on direct interactions with 'companion species' including many domesticated animals (Haraway 2007). Therefore, this research has also drawn upon the literature on human-wildlife conflict (e.g. Knight 2000a; White and Ward 2010; Redpath, Bhatia and Young 2014; Hill 2015) and cultural representations of wildlife (e.g. Buller 2004; Lorimer 2007) to understand the particular, peculiar role of the badger in British society and how this has shaped public debates over bTB.

Historical and Cultural Framings of British Badgers

We know that epidemiological links between bTB in domestic cattle and wild badger populations were not made until 1971, when a dead badger was found on a farm in Gloucestershire undergoing an anomalous outbreak of bTB (Muirhead 1972; Muirhead, Gallagher and Burn 1974). But what led the veterinarians investigating the case to make the connection to bTB and carry out a postmortem on the carcass? Furthermore, what led to a full-scale badger-culling policy being adopted by the UK government as early as 1975 (Grant 2009)? While a fuller historical investigation of this period is now underway, aimed at answering these questions in depth (Cassidy, Mason Dentinger, Schoefert and Woods, in press), examining the historical and cultural roles played by badgers has provided a productive starting point. Human languages and cultures are saturated with representations of animals that provide a rich source of information about the roles played by animals in society at a particular place and time (Corbett 2006: chap. 7).

Badgers were named and legally designated as 'vermin' in England under the Tudor Vermin Acts of 1532 and 1566, which listed those animals the Crown believed to interfere with human activities, particularly around food production, and offered financial rewards for their dead bodies. In this listing, badgers fetched a generous bounty of twelve old pence per head: a high price only shared by one other animal – the fox. In his study of churchwarden records of the payment of these bounties, Roger Lovegrove (2007: 230) intriguingly reports that despite this high reward, relatively few badgers were killed under this and later systems of vermin control, often implemented by landowners and gamekeepers. Traditionally, badgers were also eaten, and parts of their bodies were used by humans (e.g. in magical charms, hair for shaving brushes and badger fat as a liniment) (Hardy 1975; Lovegrove 2007). At least some of these practices continue in parts of continental Europe today (Griffiths and Thomas 1997; Roper 2010: 33).

This initial impression of a wildlife species with a long-standing conflict-ridden relationship with humans can be confirmed by looking at how badgers are discussed in the *London Times* newspaper's online archive, which covers 1785–1985. The earliest references found occur as part of the newspaper's routine sports reporting in the early nineteenth century:

EASTER MONDAY SPORTS
The first symptoms of sporting amusement that caught our observation appeared in the neighbourhood of Hampstead and Kentish-town. The Sun had scarcely surmounted the horizon, and – "tinged with gold the village spire," the sportsmen of Kentish-town had assembled at the Bull and Gate, to prepare for a badger hunt; and fortified by their morning draughts, they set out for the field, not with Deep-mouthed hounds, and mellow toned horn, but keen scented terriers, and high-bred bull-dogs, to assail the grizzly savage in his den, situated in a field between Highgate and Hampstead. (*The Times Archive*, issue 8271, 16 April 1811, p. 3, col. C)

While the bloodsport of badger baiting was made illegal in the United Kingdom in 1835, the related practices of digging and hunting for badgers continued throughout the nineteenth century, and discussions of these as routine and popular activities continue in *The Times* up until as late as 1911, after which the coverage shifted towards a more modern mode of disapproval and/or concern for animal welfare (see also Pease 1898).

It is not until 1877 that more positive representations of badgers started to appear, in the first of several sets of exchanges on the letters page of the newspaper:

> 'The Badger' (Letters to the Editor)
> On fine evenings we can watch them dress their fur-like coats, or do kind office for each other, and search for parasites after the manner of monkeys. No creature is more cleanly in its habits [ ]…they scrape their feet in dirty weather, and keep their house inodorous by depositing their excrement at one place for many months and covering it with earth. (Ellis, *The Times Archive*, issue 29081, 24 October 1877, p. 5, col. E)

This depiction of a clean, gentle, sociable and civilized animal was subsequently riposted in another letter, describing instead a vicious predatory animal that makes a persistent nuisance of itself to farmers:

> 'Badgers' (Letters to the Editor)
> That badgers dig out and eat young rabbits is a fact that can be documented beyond doubt in this district during the summer months to anyone who is incredulous on the subject. . . . In the early part of this year I was told by a farmer – whose veracity I have no reason to doubt – that he had been so annoyed by badgers treading down his crops in passing from one earth to another that he determined to dig them out, so that he could trap them. (Barnes, *The Times Archive*, issue 29102, 17 November 1877, p. 4, col. F)

Similar exchanges, often between amateur (and later professional) natural historians or zoologists and farmers or landowners, occurred in *The Times* every few years up until the early 1940s. As well as the above emphasis on cleanliness and sociability, in the early twentieth century people who liked badgers saw them as brave, strong, family-oriented, 'ancient' and quintessentially British in character:

> 'Men and Badgers' (Editorial)
> The badger's kin may have lived in that spot centuries before there were any human beings there. Like the best people of ancient breeding, they had kept themselves to themselves, hiding by day, coming inoffensively out by night, resisting only – and then to the death – the attempts of the upstarts and interlopers to make of them either sport or shaving-brushes. (*The Times Archive*, issue 44567; 28 April 1927, p. 15, col. E)

Badgers were also lauded for making themselves 'useful' to people in many ways, particularly by eating rabbits (a major agricultural pest at the time; see Bartrip 2008), small rodents and nuisances such as wasps' nests. They were also considered to make excellent pets. Alongside their habits of digging and crop destruction, badgers' detractors also accused them of taking ground-nesting birds (including pheasants), chickens and even young lambs. Badgers were also considered to be a problem due to their perceived interference with foxes and by extension the elite (and economically important; see Bresalier and Worboys 2014) practice of foxhunting:

'The Badger: Damage Caused in the Hunting Countries'
(Letters to the Editor)
In a hunting country, besides adding largely to the poultry claims, he does great damage by opening earths which have already been stopped. He will take possession of foxes' earths and evict the rightful owners, in many cases driving them out of the coverts with which those earths are situated. (Lascelles, *The Times Archive*, issue 46226, 31 August 1932, p. 6, col. B)

Over time, these negative arguments became rarer, and this kind of correspondence gradually stopped. Badgers hardly ever appeared in *The Times* through the middle of part of the twentieth century, until public campaigns to legally protect the animals got into full swing during the mid-1960s, culminating in the passing of the Badgers Act of 1973. Further legislation followed in 1992, making it a serious offence to 'kill, injure or take a badger', or to damage or interfere with a badger sett without a government-issued licence (Roper 2010).

This broad trajectory, from routinely hunted vermin animal, through societal conflict over badgers, to a valued and protected wildlife species, can also be traced in British cultural representations of badgers. In particular, it can be seen in the development of the most famous of these, Mr. Badger from Kenneth Grahame's 1908 novel *The Wind in the Willows*. Grahame's book is a classic of British children's fiction: as well as undergoing multiple reprints, it has been adapted for the stage, radio and television many times and has an enduring popularity. Mr. Badger is antisocial, living in the depths of the Wild Wood: he is intelligent and wise, brave and a fierce fighter in the defence of his friends, Toad, Water Rat and Mole, whom he acts as a father figure to. Visual representations of Mr. Badger have changed over the century since the book was first published: while the earliest illustrated edition of the book, published in 1913, showed Badger and the other animals in a naturalistic style (Graham and Branscombe 1913), it is the anthropomorphized illustrations of E.H. Shepard (who also illustrated A.A. Milne's *Winnie the Pooh*) that are most widely known (Grahame and Shepard 1931). This image of Badger as country gentleman then transformed into the stern, grandfatherly, spectacle-wearing character voiced by Michael Horden in a 1980s television adaptation (Hall and Cosgrove 1984–1987), and can be seen, for example, on the cover of the current Walker Illustrated Classics edition, showing the other animals gathered around Badger in storytelling mode (Graham and Moore 2009).

These characteristics can be traced back to an Anglo-Saxon riddle poem[1] dating back to the tenth century, which tells the story of an animal that lives underground in a hill, fighting and defending his family against digging invaders (Nelson 1975). This theme was picked up by the British romantic poet John Clare during the early nineteenth century in his vividly written poem *The Badger* – written from the point of view of the animal being baited (Clare, 2008)[2] – and was reprised by the First World War poet Edward Thomas in *The Combe* (1917). Following *The Wind in the Willows*, a more academic, less heroic version of the character appears in T.H. White's

The Sword in the Stone (1938), a children's novel about the young King Arthur, while more martial versions can be seen in C.S. Lewis's Narnia novel *Prince Caspian* (1951), and fantasy author Brian Jacques's *Redwall* series (1986–2010; e.g. Jacques 1988). Colin Dann's *Animals of Farthing Wood* series (1979–1994; e.g. Dann 1979) reinterpreted the character with a more environmental angle: Badger became a leader of a group of animals evicted from their home woodland by a housing development; this series was also turned into a children's animation during the 1980s. While environmental politics became more explicit at this time, these sources all share a common theme of reflection on human relationships with the British natural environment and countryside.

Another aspect of Badger's character can be seen in the association of badgers not only with British national identity in general, but also with the idea of rootedness and a specific sense of place. This can be seen in the usage of 'badger' or older, reputedly 'Celtic' versions such as 'brock' in place names (e.g. Broxbourne, see figure 4.2). Badgers also feature in real coats of arms, the fictional heraldry of House Hufflepuff in J.K. Rowling's *Harry Potter* novels (e.g. Rowling 1997) and 'heraldic' commercial imagery such as that employed by the Dorset-based Hall and Woodhouse brewery chain (brewers of 'Badger' branded beer). These kinds of deep-rooted connections between animals, place, landscape and British national culture have also been documented for several other wildlife species, including foxes (Marvin 2001), otters (Syse 2013) and wild birds (Moss 2004; Cammack, Convery and Prince 2011).

Badgers have also played a significant role in histories of popular natural history and zoology in the United Kingdom from the mid-twentieth century onwards. *The Badger* (1948) was one of the earliest publications in the influential Collins *New Naturalists* book series (Neal 1948; Marren 1995). As the write-up of one of the first systematic field studies of badgers, *The Badger* acted simultaneously as a major scientific monograph as well as an exceptionally popular natural history book of the postwar period. It was republished as a mass-print Pelican paperback in 1958 and subsequently stayed in print until 1977. Its author, Ernest Neal, worked as a schoolmaster, but was also an amateur naturalist and pioneer of nature photography, capturing the earliest still colour and video film footage of wild badgers, and eventually

Figure 4.2. Traditional coat of arms and modern corporate logo of the UK borough of Broxbourne, located in the southeast of the country. *Source*: reproduced by kind permission of Broxbourne Borough Council (www.broxbourne.gov.uk).

gaining his PhD in 1960.[3] Via this work, he became closely involved in the early development of the BBC Natural History Unit (Marren 1995; Neal 1994), and thereby played a role in the mutual constitution of wildlife documentary and the sciences of animal behaviour during the twentieth century (Davies 2000). He worked in collaboration with amateur natural historians and professional zoologists, having a profound influence on subsequent research in field biology (Roper 2010), and influenced policy via membership of the Badger Consultative Panel for many years (Neal 1994). *The Badger* (1948) is a compelling combination of scientific monograph, first-hand narrative and 'how-to' manual for the aspiring amateur natural historian, clearly explaining how to conduct your own field studies of badgers if you should so wish. It describes badgers as clever, sociable, clean, civilized and family-oriented animals: terms immediately familiar from the above discussions in *The Times*.

In more recent popular cultural contexts, badgers have become increasingly abstracted: they often appear as part of a revolving cast of what Tess Cosslett describes as 'human beings with animal heads' (2006: 181) in children's television and books: essentially stories about humans (usually children) who act as humans, but just happen to be depicted as animals. Particularly since the 1990s, this tendency towards abstraction has accelerated into surrealism and comedy, with even the word 'badger' being seen as simply funny, featuring in wordplay and stand-up comedy routines in the United Kingdom.[4] Badgers have even had their own Internet craze: the *BadgerBadgerBadger* animation, which simply involves animated badgers doing star jumps, followed by a mushroom and a snake, to the accompaniment of a catchy tune (Picking 2003). It is clear that in these contexts the badger's strikingly striped, monochrome face plays an important role. Indeed, visual representations of badgers often appear: from coats of arms through commercial and campaign logos, book illustrations, nature photography and documentary, visual arts, television programmes, and of course countless soft toys. The face lends itself to abstraction, and is adopted beyond badger-specific issues into broader environmental and conservation campaigning, as can be seen in this logo of the national Wildlife Trusts (figure 4.3), a conservation NGO (Non-Governmental Organization).[5]

These highly positive images and ideas stand in sharp contrast to one of the few negative portrayals of badgers in British popular culture: Beatrix Potter's *The Tale of Mr. Tod* (1912). Like Kenneth Grahame, Beatrix Potter wrote children's books in the first decades of the twentieth century, similarly

Figure 4.3. Logo of the UK Wildlife Trusts. *Source*: reproduced by kind permission of the Wildlife Trusts.

featuring humanized, clothed animals adventuring across a specifically English countryside. Unlike Grahame, whose work tended towards the allegorical and surreal, Potter's many animal tales were intended to teach children basic biology and natural history (Cosslett 2006). Famously, Potter even upbraided Grahame for his lack of realism when portraying animals: 'Did he not describe Toad as combing his hair? A mistake to fly in the face of nature – A frog may wear galoshes; but I don't hold with toads having beards or wigs!' (Potter 1942, in Potter and Taylor 2012).

To modern eyes, *The Tale of Mr. Tod* is a grim (but very biological) tale of predation: the badger, Tommy Brock, uses guile to kidnap a nest of baby rabbits, which are then in turn stolen from him by the fox, Mr. Tod. The two fight, and in the confusion the babies' relatives (Peter Rabbit and Benjamin Bunny) manage to rescue them. While Potter's claims to realism are clearly somewhat overstated, badgers do

Figure 4.4. 'Tommy Brock', illustration from Beatrix Potter's *The Tale of Mr. Tod* (1912: 21)
Source: copyright © Frederick Warne & Co. 1912, 2002. *Source:* reproduced by permission of Frederick Warne & Co.

indeed predate on young rabbits by digging them out of their protective burrows. While this behaviour is well-known by biologists (e.g. Roper 2010: 116–120), it caused a minor sensation when accidentally recorded by BBC cameramen during the 'reality TV' natural history series *Springwatch* several years ago (Scoones 2009). Potter's Tommy Brock is a deeply unpleasant character, who as well as being sly and predatory is smelly, dirty, uncouth and carries a spade (much like baiters and diggers), as seen in figure 4.4. He is also portrayed as an agricultural labourer in contrast to the 'gentleman' Mr. Tod, the fox. This reflects the class politics of animal hunting in the United Kingdom. While foxhunting was developed by the landed gentry during the nineteenth century into an iconic feature of British country life (Marvin 2001) and otter hunting was associated with a broader spread of participants including the middle classes (Allan 2010), with some overlaps badger hunting and baiting in particular were and still are associated with working-class people (Griffin 2007: 84–85).

Framings of Badgers in Contemporary bTB Debates

Having surveyed the roles played by badgers in British historical and cultural sources, we now return to the contemporary bTB controversy. How have badgers been framed in mass media coverage and contribution to these debates, and are there any commonalities or continuities to be found with earlier representations of the animals? The principle data source for this analysis has been the LexisNexis UK national newspaper online archive, which was searched over the period 1995–2010 for 'badger' and 'TB'. This produced a core sample of newspaper articles, which were analysed using grounded theory (e.g. Strauss and Corbin 1998) – an iterative process of questioning, reading and qualitative coding, which continued until the analysis stabilized. The core sample was then supplemented with material from online sources, social media, specialist agricultural press, parliamentary proceedings, government publications and relevant TV and radio programmes. Framing analysis, a widely used methodology for media analysis, was also employed to investigate how actors involved in the debate have understood the problem and what should be done about it. This coverage has had a very tight focus on badgers and the associated question of whether they should be culled, with less attention paid to other factors involved in the bTB issue such as cattle movement, farming practices or bTB testing regimes. The issue was also framed in one of two ways: either bTB as a chronic agricultural problem, or badger culling as a potential environmental risk. For many British people, it is likely that their first (or only) encounter with this famously nocturnal and retiring animal is via the fictional and popular cultural sources described above. Fictional badgers have therefore played a central role in this process by providing journalists with a series of easily recognizable 'hooks' from which a complex and relatively obscure science/policy issue could be discussed without losing audiences' (and editors') interest in the story. These contemporary representations take the form of 'good' and 'bad' badgers, broadly corresponding with the environmental and agricultural framings of the bTB problem; but also displaying some striking continuities with the historical and cultural sources described above.

'Good Badger'

While *The Wind in the Willows* was not the only cultural source to be referenced, it was by far the most prominent and frequently mentioned one. The book's status as a touchstone for the bTB controversy is illustrated well by a satirical cartoon in Private Eye magazine published when the Coalition government announced their culling policy. Drawn in the style of the Shepard illustrations, it depicts Toad, Mole and Ratty with guns pointed at Mr. Badger, with the caption "It's nothing personal Badger, it's just we've been paid to cull you" (Longstaff, 2010). *The Wind in the Willows* and/or 'Mr. Badger' were referenced, directly or indirectly, during discussions of the good features of badgers and/or why they should not be culled. These 'good badgers' were often described as being mysterious/shy/averse to (human) social contact; intelligent/wise; and a brave or strong fighter when attacked:

> SUPER FURRY ANIMAL OR CATTLE-KILLING, TB-RIDDEN VERMIN?
> People don't just love them because they are cuddly, but because they are so full of mystery. You see the size of their claws, and their teeth, and how quick they are, you don't want to mess with them – I suppose that was the challenge for badger baiters. (Adams, *Observer Magazine,* 4 May 2008, p. 24)

When people do encounter badgers in the wild, this seems to confer a special sense of connection with wildlife and the natural world:

> IF YOU GO DOWN TO THE WOODS TODAY
> The first time a young badger bounced down the garden to greet me, I felt a flush of pride. Presumably it had mistaken me in the dark for a fellow badger – it bolted the second it realised its error. But it was still gratifying, as if its snuffling at my feet conferred some kind of seal of approval from the natural world. (Askwith, *The Independent,* 13 June 2003)

The relative rarity for some people of experiencing such connections may in part account for the popularity of 'badger watching' as a leisure activity in the United Kingdom.[6] While people rarely encounter badgers by accident, due to their nocturnal lifestyle, a combination of poor eyesight, routine foraging habits and an omnivorous diet means that given the right conditions it can be relatively easy for an amateur to seek out and observe them.[7] This happens along a continuum with people feeding badgers visiting their gardens on an ad hoc basis at one end, and organized holiday breaks at the other:

> TEA FOR TWO, PLEASE
> 'SAT under a beech tree in the cow field behind the cottage and waited,' reads an entry in the visitors' book at Westley Farm in the Cotswolds. 'At about 9.20pm a large badger came across the field towards me. He came up

to the tree and sniffed around the roots on which I was sitting. Wow! Then something I will never forget – he sniffed at the leg of my jeans, stopped, looked up, our eyes met.' (Ellis, *The Times (Weekend)*, 6 July 2002, p. 6)

While initially it seems contradictory for dairy farmers (often hard-pressed due to bTB) turning to people's love of badgers as an alternative income source, this coverage suggests that farming attitudes to the badger/bTB situation may be more variable than the stereotype:

TEA FOR TWO, PLEASE
There is irony in the badger becoming a farm's best friend, particularly in Gloucestershire, where the incidence of bovine TB has been relatively high. 'Many farmers are terrified of TB and are anti-badger,' Julian says. 'But not the small farms in this valley.' The Usbornes believe that wider issues, such as feed, animal husbandry and cattle controls, need to be explored. Why, they ask, are some herds free of TB while others in the same area are infected? (Ellis, *The Times (Weekend)*, 6 July 2002, p. 606)

As described above in relation to their appearances in place names and heraldry, badgers' habit of living in family groups in one place over the long term in the United Kingdom is often linked to descriptions of 'good badgers'. These animals are still seen as holding intimate and long-standing connections with the land; also as a 'native' species they are seen as ancient residents of the country, and are highly symbolic of British national identity:

LEADING ARTICLE: IN PRAISE OF . . . BADGERS
'The most ancient Briton of English beasts,' wrote the poet Edward Thomas of the badger, a justified verdict on a black-and-white creature that has always added colour to the nation's life. The appearance of one sett in the Domesday Book merely marks the start of the current chapter in a tale stretching back a quarter of a million years. Despite their elusive nature, their inquisitive face is still one of the most recognisable symbols of British wildlife. (*The Guardian*, 15 May 2007, p. 28)

This is often underlined by pointing out connections with place names, terms such as 'brock' and 'Briton', and attributions of age and gender ('Mr.' or 'Old' Brock). As the above quote suggests, this national symbolism is closely connected to ideas about a specifically British form of wildlife/countryside/nature. Despite their clear coexistence in human spaces (farms, gardens), 'good badgers' tend to be discussed as if they spend most of their time in the 'natural' space of the Wild Wood.

An overlapping, but distinct set of characteristics involves discussions of badgers as sociable (with other badgers), family oriented and with evocatively humanlike characteristics[8]:

TB or Not TB?
They have young ones to feed at this time of year, ablutions to perform and family grooming duties, as well as house cleaning and repairs to do. It is a furiously hectic life below ground. In fact, on still frosty nights you can see a plume of steam lifting from an air vent built by the badgers at the back of their sett, like the warm white billows drifting upwards from sidewalks above the New York subway. And these animals are our deadly enemy? (Mitchell, *Daily Telegraph (Weekend)*, 4 Mar 2006, p. 1)

Although badgers are omnivores, whose diet includes small mammals (including baby rabbits and hedgehogs), birds and eggs, 'good badger' framings tend to emphasize the more innocuous aspects of this diet such as worms, insects, snails and nuts:

Black and White and Bred All Over
On a good night, badgers suck worms out of the ground just like children eat spaghetti. (Beardsall, *Daily Telegraph*, 9 August 2003, p. 12)

Unsurprisingly, framings of badgers as victims (of humans) have been particularly prominent in this coverage, directly referencing the continuing practices of baiting and digging. At times, the 'badger as victim' framing places these practices into a broader context of historical relations between humans and wildlife in general:

Stop Picking on Mr Brock:
It's the Silly Cow with TB You Should Be Blaming
Death is always the soft option – at least, it is for those not doing the actual dying. The badger cull is all of a piece with the slaughter of predators that was all the rage in the nineteenth century and still continues in some places, illegally, today. When in doubt, blame a wild creature; and then kill it. Job done. (Barnes, *The Times*, 7 October 2006, p. 23)

Finally, the badger as victim is discussed in unambiguously human terms, placing the reader directly into the shoes (paws?) of animals facing death from impersonal authorities. This is particularly clear in the repeated use of headlines such as 'The Culling Fields' (e.g. Brown 2005: 30; Fricker 2015: 25), descriptions of culling as 'mass slaughter' and references to past culls and gassing: immediately evoking the (human) horrors and holocausts of the twentieth century:

TB or Not TB?
There must be an immediate 'blitz cull' of as many as 100,000 badgers – massed gassings, total elimination zones extending across swathes of Devon, Somerset, Cornwall and any other seriously affected county. No more waiting. Push the red button. Do it now. (Mitchell, *Daily Telegraph (Weekend)*, 4 March 2006, p. 1)

This 'victim' mode is underlined by depictions of badgers as female, as children, or by likening their situation to that of refugees:

> MOTHER BADGERS ARE SNARED IN RUSTY CAGES, PARTED FROM
> THEIR SCREAMING CUBS AND COLDLY SHOT IN THE HEAD . . .
> ALL WITH THE GOVERNMENT'S BLESSING
> TRAPPED in a small, rusting cage this despairing badger paws at the bars and pushes her snout through the bars as she struggles to escape. Somewhere in the darkness a cub screams for its mother. (Weathers, *Daily Mail*, 3 June 2003, p. 11)

> A CULL BY ROYAL APPOINTMENT
> A cull – by its nature incomplete – 'would have a profound effect on the lifestyle of survivors. It might well cause changes in their immune systems which make them less resistant to disease. With their society in turmoil, bereaved badgers would almost certainly traverse the country far and wide, infecting more badgers and more cattle.'(Hattersley, *The Times*, 22 December 2005, p. 18)

This set of discourses, framing badger culling in terms of human war and its victims, provides an interesting appropriation and reversal of a set of widely used metaphors of 'the battle against disease' (Nerlich and James 2009). Rather than utilizing the bellicose language of waging war, it instead invokes anti-war rhetoric to draw attention to the potential consequences for badgers of 'eradicating' bTB.

'Bad Badger'

Given the prevalence of positive cultural representations compared to the single negative example of *The Tale of Mr. Tod*, it is striking that in contemporary media coverage of badger/bTB, negative framings of badgers are not just prominent but considerably outnumber discussions of the 'good badger'. Most obviously, 'bad badgers' spread disease by transmitting bTB to domestic cattle, and potentially to humans:

> BADGER THE GOVERNMENT
> Until it was brought under control in the 1960s, TB was a serious danger in Britain. So there is cause for concern that, in the past decade, cases of this disease in cattle have soared. An expanding badger population is blamed. These animals carry TB though they themselves appear unaffected by it. One mouthful of grass on which a diseased badger has urinated is believed to carry sufficient tubercle bacilli to infect a cow. The worry is that humans could contract the illness through consuming unpasteurised milk products. (*The Times*, 5 May 1999)

In contrast to the human-like 'good badger', 'bad badgers' tend to be discussed in the plural, and in a depersonalized way. This plays into discussion of links between rising badger populations and the spread of disease (Veterinary Association for Wildlife Management 2010).

In turn, this forms part of a complex of characteristics depicting badgers as an undesirable underclass: violent, disruptive, criminal and far too numerous:

> INSIDE STORY: WHAT HAVE I DONE TO DESERVE THIS?
> In the book [*The Wind in the Willows*], Badger is a solitary creature. Round here, he'll be shacked up with a dozen friends and family. And quite likely, he'll have tuberculosis. (Perkins, *The Guardian*, 12 April 1999, p. 8)

> SUPER FURRY ANIMAL OR CATTLE-KILLING, TB-RIDDEN VERMIN?
> As the badger population has grown, they are increasingly in our back yards; as Colin Gray points out, just as we had urban foxes, increasingly we will be seeing urban badgers as they travel further in search of food. In some places this is already a reality. In Evesham last year, 'a rogue badger attacked five people during a 48-hour rampage in a quiet suburb'. In one suburb in Sheffield, it was recently reported, residents 'were demanding an Asbo for sex-mad badgers'. There were 19 setts in a hundred yards of back gardens. Michael Broomhead, 60, a retired butcher, said: 'They have felled three trees by digging under them. When they are having sex they howl and scream, and when they are fighting they make terrible bloodcurdling noises as if they are being murdered.' (Adams, *Observer Magazine*, 4 May 2008, p. 24)

While violence, or the capacity for violence when under attack, is present in the 'good badger', here we see badgers' predatory and violent behaviour being greatly accentuated:

> CULL OR CURE DILEMMA AS BADGERS GET BLAME FOR EPIDEMIC
> Ground-nesting birds have also suffered in the explosion of badger numbers, according to Mr Barker [a dairy farmer]. He believes there could be up to 50 badgers in the main sett alone. 'I now have no lapwings, curlews or wild pheasants because there are so many badgers searching for food and taking all the eggs in the spring.' (Goodwin, *The Independent*, 1 December 1997, p. 20)

Other negative aspects (or those disruptive to humans) of badger behaviour are also emphasized, such as crop destruction and digging. Badgers live in underground setts, continually dig and are opportunistic, intelligent omnivores: therefore, such activities at times bring them into conflict with humans (Roper 2010: 267–298):

THOUSANDS OF BADGERS ARE CONDEMNED TO DIE OVER TB FEARS
Farmers detest the nocturnal mammals not only because of the belief that they spread TB but because they flatten cereal crops, nibble growing corn on the cob, and even strip vineyards of grapes. (Hinsliff, *Daily Mail*, 18 August 1998, p. 21)

THE SMART SETT
Aberdeenshire council is spending pounds 30,000 on a new council home for a family of badgers because their present sett has undermined a main road between Huntly and Banff. (*Daily Telegraph*, 21 March 2001, p. 25)

In the 'bad badger' framing, these features come together to depict an agent of chaos – and, due to legal protections, one that can escape 'justice':

A VERMINOUS VIETCONG STALKS THE COUNTRYSIDE
Not since the Beast of Bodmin, not since the Hound of the Baskervilles, had so awful a creature plagued the countryside. *Meles vulgaris*, something between a weasel and a bear, was overrunning hill and dale. And it was, of course, Labour's fault. What were the teddy-hugging, town-dwelling, pizza-eating classes going to do about it, I was asked? They would not be content until every rustic parlour was a zoo of free-range foxes, badgers, stags, kites and predatory geese? I could not argue the damage. Across the landscape meadows were being upheaved, hedges, banks and bridle-ways subsiding, tennis courts falling into holes. Tunnels of Ho Chi Minh ingenuity were sapping the ancient walls and lawns of England with a verminous Vietcong. These omnivorous monsters were eating lambs and ground-nesting birds. They were the only known predator of the hedgehog. Archaeological sites were being destroyed. The killer brock was prowling at will, cockily secure under the 1992 Protection of Badgers Act. (Jenkins, *The Times*, 4 April 2004, p. 24)

While this piece, in common with much newspaper commentary, is clearly meant to be humorous, it still expresses a common frustration with the legal protection of badgers in the United Kingdom. It also illustrates how controversies over badger culling are intertwined with tensions between traditional British rural centres of power and modern urban elites – at the time epitomized by successive Labour administrations. During their time in power (1997–2010), as well as gradually withdrawing licences for farmers to cull badgers, and ruling out a culling policy in 2008, the Labour government also outlawed foxhunting in 2005 (Woods 2008). While the specifics of foxhunting controversies are distinct to those surrounding badger culling and baiting/digging (Marvin 2000), current intersections of power and political interests result in their present alignment (see, e.g. May 2010).

These 'good' and 'bad' badger framings are employed strategically by media and other actors engaged in arguments for and against culling policies. However, as is suggested by some of the examples above, each trope is also employed, albeit in an exaggerated form by the 'other' side as well, for example when pro-cull actors cite *The Wind in the Willows* as the source of 'emotional' popular resistance to culling (Tasker 2012). Indeed, in some longer articles, authors switch rapidly between the two tropes, using them as a resource to explore the issue at depth. Wyn Grant (2009) and Gareth Enticott (2011b) also identify the use of good/bad badger tropes, respectively by policy makers during the 1960s and 1970s and farmers currently affected by bTB carrying out illegal culling. However, these take a slightly different form, involving the attribution of bTB infection to specific individuals: sick 'rogue badgers' whose behaviour is abnormal in many ways and must be 'taken care of' (Enticott 2011b: 204; Maye et al. 2014: see also Jenkins 2004, quoted above). This trope was rarely present in national press coverage of badger/bTB, although the related idea of expressing concern for the suffering of sick badgers was employed as part of pro-cull rhetoric:

> ANIMAL LOBBY CONDEMNS BADGERS TO SLOW DEATH
> It [bTB] is also causing great suffering to the badgers themselves. Thousands die each week from the long-drawn-out effects of the disease (unless, as any West Country roadside bears witness, they are so weakened that they fall victim to a passing vehicle). (Brooker, *The Daily Telegraph*, 7 March 2005, p.14)

This illustrates the flexibility of good/bad badger discourses, and the ways in which they can be taken up and strategically reshaped to suit changing contexts and audiences. While the link with TB is new, the association of the 'bad badger' with dirt and solitary, violent behaviour is evident in Beatrix Potter's Tommy Brock, created nearly one hundred years earlier.

Discussion

This chapter has traced how badgers have been represented in British society via an analysis of historical and cultural sources, alongside media coverage of contemporary debates over bovine TB and badger culling. Across all these sources, two opposing characters dominate: the 'bad badger' and the 'good badger', which today are broadly associated with arguments for and against culling policies for bTB management. This analysis suggests that badgers have occupied an ambivalent position in British society since at least the mid-nineteenth century, and that their connection with bovine TB has intensified this conflicted role since 1971. Various aspects of badger behaviour bring these animals into conflict with humans: on the other hand, they have a range of other characteristics that people have found admirable and aesthetically pleasing. Some of these, such as strength, bravery and loyalty to family, are features that may be displayed during baiting and digging activities: this valorization of and sense of closeness to the 'hunted' animal by the 'hunter' is not unusual (Marvin 2000; Carvalhedo Reis 2009). Through long-standing British cultural traditions surrounding nature and the countryside, as well as in more recent traditions of popular natural history, badgers are widely regarded as 'charismatic animals' in the United Kingdom (Lorimer 2007). Charismatic wildlife species such as the panda are often used in conservation campaigning: the fact that in the United Kingdom badgers take on such a role despite not being an endangered species attests to the strength of this status, and may go some way towards explaining the strength of opposition to badger culling (Bennett and Willis 2008; DEFRA 2011).

At the same time, many of the characteristics of the 'bad badger' highlight how these animals can come into conflict with humans. Alongside the obvious issue of bTB transmission, these include crop-raiding, predation, violence and disruptive digging. Such descriptions are congruent with the kinds of language used to describe animals in conflict with humans across a wide variety of species, cultures and locales: such animals are often ascribed to the category 'pest' or 'vermin'. John Knight's (2000b) framework of 'pestilence discourses', in which pest animals tend to be represented as dirty, violent, criminal, cunning, numerous and out of control, while their harmful effects upon humans are emphasized, fits well with these negative framings of badgers. This is also apparent in the transition from debates about badgers to debates about badgers and bTB. The historian Mary Fissell (1999) has argued that the category 'vermin' in early modern England related to animals in direct competition with humans for resources, which could explain why badgers and other vermin animals were (and are) frequently portrayed (and treated as) criminals (see, e.g. Cassidy and Mills 2012). Fissell argues that early modern 'vermin' were not associated with dirt, disease and disgust as they are today, and that these links later developed alongside the adoption of germ theory during the nineteenth century (Douglas 1966). Lucinda Cole (2014) draws upon sixteenth- and seventeenth-century accounts of the plague to instead argue that vermin *were* associated with disease at this time, but via an intertwined complex of natural and supernatural causes including miasma (bad air) and witchcraft.

In either case, badgers were not directly connected with bovine TB until the early 1970s; and as we have seen, historical and contemporary representations of the 'bad badger' tend to focus disproportionately on their roles as agricultural pests, rather than as disease carriers. Therefore, contemporary associations between badgers and bovine TB appear to be facilitated by *both* historical understandings of badgers as animals in competition and conflict with humans, and by modern scientific understandings of them as disease vectors. Research on societal representations of disease supports this idea: we see a common language of war, apocalypse, the enclosure of safe space (biosecurity) and attributions of risk outside such spaces across cases including foot-and-mouth disease, 'superbugs', pandemic influenza and bTB (Nerlich and James 2009; Nerlich, Brown and Wright 2009; Washer 2010). This includes the 'othering', dehumanization and exclusion of groups and individuals seen as the source of disease risks (Joffe 1999). In this case, badgers have been treated in much the same way, transferring societal anxieties about disease risks onto an animal species: these associations between 'vermin', disease anxieties, prejudice and excluded groups have been applied to humans and other animals alike (Marcu, Lyons and Hegarty 2007; Mavhunga 2011). It is possible that the badger/bTB controversy is part of a broader escalation of risk narratives around animals in European media and popular culture in recent years (Gerber, Burton-Jeangros and Dubied 2011).

Knight (2000b) also highlights the dualistic nature of pestilence discourses. This is very much in evidence in the material presented here, as are the ways in which 'pest' animals tend to disrupt the spatial, bodily or psychological boundaries constantly being constructed between humans and other animals, as well as 'culture' and 'nature'. As described above, these framings of badgers invoke important notions of space, and the proper occupation of space. The 'good badger' is symbolic of an idealized British nature or countryside, and as such is often depicted occupying 'natural', non-human spaces such as Kenneth Grahame's 'Wild Wood'. By contrast, 'bad badgers' are invariably framed as intruders into human, albeit agricultural, spaces where they disrupt and impede human activities (see also Spencer 2010). The constructed nature of these boundaries, and the way in which they must constantly be (re)negotiated (Schlich, Mykhalovskiy and Rock 2009) means that the 'pest' role is inherently ambiguous. This explains why constructions of pests tend to flip between positive and negative versions, particularly when such boundaries are being contested (e.g. Hytten 2009; Potts 2009; Brown 2011; Cassidy and Mills 2012).

By comparing contemporary public debates over badgers and bTB with older representations of badgers in the United Kingdom, a tentative historical narrative can be traced. From an early period in which badgers were legally considered to be vermin and hunted (although also admired), during the late nineteenth and early twentieth centuries the 'good badger' gradually appeared and become more prominent. During this period, the emergence of early animal welfare movements and resulting changes in social attitudes marginalized, and eventually outlawed many hostile interactions with badgers via baiting and formal hunting (Griffin 2007). The first two decades of the twentieth century seem to have been a turning point of sorts: while it may be a coincidence that *The Wind in the Willows*, *The Combe* and *Tales of Mr. Tod* were all published within ten years of each other, the British anti-bloodsports movement also came to the fore at this time (Allen 2010: 81; Griffin 2007). Through the twentieth century positive representations of badgers became increasingly common, particularly once they had featured in BBC wildlife programming (e.g. Neal and Hewer 1954; Bale 1977). By the late twentieth century the 'good badger' had become so dominant that it became increasingly abstracted and even parodied by comedians. This most recent incarnation, the 'surreal badger', has been adeptly mobilized by contemporary anti-culling campaigners, transforming into a symbol of resistance to the Conservative government that came to power in 2010 (figure 4.5).[9]

Today's controversy over badgers and bTB can be read as a continuation of the long-standing 'badger debate' about the appropriate position of these animals in British society. Are badgers pests to be 'managed' and removed when they get in our way; or a cherished, charismatic wildlife species to be preserved and protected?

Figure 4.5. 'Bring It'; *Daily Mash* badger T-shirt design *Source*: reproduced by kind permission of Paul Stokes / *Daily Mash*.

Contemporary associations between disease, risk and animals (particularly pests) also appear to have contributed to the re-emergence of the 'bad badger' into public discourse in the United Kingdom.[10] These deep-rooted connections suggest that a deeper understanding of many other contemporary conflicts between humans and wildlife can be gained by investigating the historical development of such conflicts, and how cultural representations of the animals involved have changed over time.

These findings have important implications for current debates over the management of bovine TB in the United Kingdom. Firstly, there is a pressing need to reframe the controversy beyond the reductive yes/no question of badger culling, and to 'open up' (Stirling 2008; Leach, Scoones and Stirling 2010) and investigate the broader questions of the underlying and highly complex problem of bTB spread, its potential causes and what can and should be done about it. A key first step would be to acknowledge the existence of the historical and contemporary 'badger debate' underlying the bTB controversy. This would enable researchers and policy makers to mobilize human-wildlife conflict frameworks to investigate and attempt to mitigate problems of human/badger coexistence that are *unrelated* to bTB. Such a recognition could start to decouple public debates about badger protection, conservation and coexistence from those about bTB control, farming and the rural economy.

Secondly, as research on the disease ecology of bTB is increasingly recognizing (McDonald 2014; O'Connor et al. 2012; Byrne et al. 2012; Broughan et al. 2016), human cultures, histories, politics, ethics, economics and actions play important roles in maintaining, spreading and managing the disease across multiple species. To improve our understanding of how bovine TB works, and how it might best be managed, the multiple social and biological causes of this chronic disease problem must be further investigated in concert, using a broader range of qualitative and quantitative research methodologies. Only then can we hope to work towards some kind of sustainable resolution of this protracted and divisive controversy, which has yet to find any satisfactory solutions for the humans and other animals involved.

Acknowledgements

A version of this chapter was originally published in 2012 as 'Vermin, Victims and Disease: UK framings of badgers in and beyond the bovine TB controversy', *Sociologia Ruralis* 52(2): 192–204. Research was initially carried out with the support of an Interdisciplinary Early Career Fellowship from the UK Rural Economy and Land Use Programme (grant no. RES-229-27-0007-A), and further developed with the support of a Medical Humanities Fellowship from Wellcome Trust (101540/Z/13/Z). I would like to acknowledge the contribution of Charlotte Kenten to the early stages of this research, and thank my various fellowship mentors, particularly Abigail Woods, Peter Simmons, Graeme Medley and Robbie McDonald, for their feedback and support in developing this work.

Angela Cassidy is a Lecturer in the Land, Environment, Economics and Policy Institute, Department of Politics, University of Exeter. She is currently undertaking a Wellcome Research Fellowship investigating the history of bovine tuberculosis in the United Kingdom since the late 1960s and the long-standing public controversy over whether to cull wild badgers in order to manage the disease in domestic cattle herds. Her research spans the contemporary history of science and medicine, science communication and science and technology studies: she has particular interests in public scientific controversies; interdisciplinarity; environments, food and agriculture; and nature/culture. Her earlier work has investigated these issues in case studies of 'One Health' (advocacy for the convergence of human and animal health), food-chain risks and public debates over popular evolutionary psychology.

Notes

1. The nature of the poem is such that the animal is never identified, but it has popularly been considered to be a badger: for arguments to the contrary, see Bitterli (2007).
2. Alongside other Romantic poets, Clare's work was mutually shaped by the incipient animal rights movement in Britain; see Perkins (2003) for a more in-depth discussion.
3. The cover of Neal's autobiography (1994) simultaneously references his own work and the cultural role of the badger by depicting the animal in schoolmaster's robes.
4. See, e.g. material by the British comedians Eddie Izzard, Marcus Brigstocke and Harry Hill.
5. I understand that part of the reason for the adoption of the badger face as the logo for the UK Wildlife Trusts is that the simple monochrome image could be easily copied using the scarce resources available to NGO campaigners at the time (Owain Jones, pers. comm.); see also Nicholls (2011) for the role of visual abstraction of animal images in conservation logos.
6. Many of the UK Wildlife Trusts, as well as local landowners and farmers, now conduct supervised or unsupervised 'badger watching' on a routine and frequently commercial basis; see, e.g. http://www.wildlifetrusts.org/stake-out-a-badger-sett.
7. This contrasts with the notorious difficulties of 'seeing' or indeed studying many wild animals (Rees 2006). Similarly, the relative visibility of birds plays an important role in the popularity of bird watching as a leisure activity (see Cammack et al. 2011; Law and Lynch 1988; Moss 2004).
8. I generally avoid the term 'anthropomorphic', as it seems too normatively loaded to be helpful in understanding how and why people tend to highlight the similarities between themselves and other animals in this way; see Daston and Mitman (2005) for further discussion of this issue.
9. In 2013 the rock star Brian May (funder of the anti-cull *Save Me* campaign) teamed up with the makers of *BadgerBadgerBadger* to create a Save the Badger web animation (Picking, May and Blessed 2013). See also the Badger Penalty Shootout online game responding to then Environment Secretary Owen Paterson's comments that 'the badgers have moved the goalposts' (Political Scrapbook 2013), and satirical commentary on the situation (e.g. Daily Mash 2012, 2014).
10. While it is far beyond the scope of this article to draw direct links between this re-emergence and changes in human attitudes or practices towards badgers, Enticott (2011b) suggests that the severity of the bTB problem (and farmers' lack of power to prevent it) may be leading to an increase in illicit persecution and killing of badgers.

References

Adams, T. 2008. 'Super Furry Animal or Cattle-Killing, TB-ridden Vermin?' *Observer Magazine*, 4 May, p. 24.
Allen, D. 2010. *Otter*. London: Reaktion Books.
Askwith, J. 2003. 'If You Go Down To The Woods Today. To Some, they're Charming, Gentle Creatures. To Others, they're a Menace', *The Independent*, 13 June.
Atkins, P.J. 2000. 'Milk Consumption and Tuberculosis in Britain, 1850–1950', in A. Fenton (ed.), *Order and Disorder: The Health Implications of Eating and Drinking in the Nineteenth and Twentieth Centuries*. East Linton: Tuckwell Press, pp. 83–95.

Atkins, P.J., and P.A. Robinson. 2013a. 'Coalition Culls and Zoonotic Ontologies', *Environment and Planning A* 45(6): 1372–1386.

———. 2013b. 'Bovine Tuberculosis and Badgers in Britain: Relevance of the Past', *Epidemiology and Infection* 141(7): 1437–1444.

Bale, P. (producer). 1977. *Badger Watch*, BBC One, British Broadcasting Corporation. http://www.wildfilmhistory.org/film/288/Badger+Watch.html.

Barnes, G. 1877. 'Badgers' (Letters to the Editor), *The Times Archive*, issue 29102, 17 November, p. 4, col F.

Barnes, J. 2006. 'Stop Picking on Mr Brock: It's the Silly Cow with TB you Should be Blaming', *The Times*, 7 October, p. 23.

Bartrip, P.W.J. 2008. *Myxomatosis: A History of Pest Control and the Rabbit*. London: I.B.Tauris.

Beardsall, J. 2003. 'Black and White and Bred All Over: Thanks to Dairy Farming, Badgers are Thriving in Areas such as Somerset, says Jonny Beardsall', *Daily Telegraph*, 9 August, p. 12.

Bennett, R.M., and K.G. Willis. 2008. 'Public Values for Badgers, Bovine TB reduction and Management Strategies', *Journal of Environmental Planning and Management* 51(2): 511–523.

Bitterli, D. 2007. 'Exeter book Riddle 15: Some Points for the Porcupine', *Anglia - Zeitschrift für englische Philologie* 120(4): 461–487.

Boyd, I.L. 2015. 'Bovine Tuberculosis: DEFRA Responds to Badger-Cull Critique', *Nature* 527 (7576): 38. doi:10.1038/527038b.

Bresalier, M., and M. Worboys. 2014. 'Saving the Lives of our Dogs: The Development of Canine Distemper Vaccine in Interwar Britain', *The British Journal for the History of Science* 47(2): 305–334.

Brooker, C. 2005. 'Animal Lobby Condemns Badgers to Slow Death. TB EPIDEMIC. Notebook', *The Daily Telegraph*, 7 March, p. 14.

Broughan, J.M., et al. 2016. 'Farm Characteristics and Farmer Perceptions Associated with Bovine Tuberculosis Incidents in Areas of Emerging Endemic Spread,' *Preventive Veterinary Medicine* May. doi:10.1016/j.prevetmed.2016.05.007.

Brown, J. 2005. 'THE CULLING FIELDS: Mass slaughter plan for badgers provokes outcry; Animal groups deny that cull will ease TB "crisis"', *The Independent*, December 16, p. 1–2.

Brown, K. 2011. 'Rabid Epidemiologies: The Emergence and Resurgence of Rabies in Twentieth Century South Africa', *Journal of the History of Biology* 44(1): 81–101.

Buller, H. 2004. 'Where the Wild Things Are: The Evolving Iconography of Rural Fauna', *Journal of Rural Studies* 20(2): 131–141.

Buller, H.J. 2013. 'Animal Geographies I', *Progress in Human Geography* 38(2): 308–318.

———. 2015. 'Animal geographies II: Methods', *Progress in Human Geography* 39(3): 37–84.

Byrne, A., et al. 2012. 'The Ecology of the European Badger (*Meles meles*) in Ireland: A Review', *Biology and Environment: Proceedings of the Royal Irish Academy* 112B: 69–96.

Cammack, P.J., I. Convery and H. Prince. 2011. 'Gardens and Birdwatching: Recreation, Environmental Management and Human–Nature Interaction in an Everyday Location', *Area* 43: 314–319.

Carvalhedo Reis, A. 2009. 'More than the Kill: Hunters' Relationships with Landscape and Prey', *Current Issues in Tourism* 12(5): 573–587.

Cassidy, A. 2006. 'Evolutionary Psychology as Public Science and Boundary Work', *Public Understanding of Science* 15(2): 175–205.

———. 2015. '"Big Science"' in the Field: Experimenting with Badgers and Bovine TB, 1995–2015', *History and Philosophy of the Life Sciences* 37(3): 305–205. doi:10.1007/s40656-015-0072-z.

Cassidy, A., and B. Mills. 2012. '"Fox Tots Attack Shock": Urban Foxes, Mass Media and Boundary-Breaching', *Environmental Communication: A Journal of Nature and Culture* 6(4): 494–511.

Cassidy, A., Mason Dentinger, R., Schoefert, K. and Woods, A. 'Animal Roles and Traces in the History of Medicine' *BJHS: Themes*, Vol. 2, in press, 2017

Clare, J. 2008. *John Clare: Major Works*, E. Robinson and D. Powell, eds. Oxford: Oxford University Press, pp. 246–247.

Cole, L. 2014. 'Of Mice and Moisture: Rats, Witches, Miasma and Early Modern Theories of Contagion', *Journal for Early Modern Cultural Studies* 10(2): 65–84.

Corbett, J.B. 2006. *Communicating Nature: How We Create and Understand Environmental Messages*. London: Island Press.

Cosslett, T. 2006. *Talking Animals in British Children's Fiction, 1786–1914*. Aldershot: Ashgate Publishing.

Daily Telegraph. 2001. 'The Smart Sett', 21 March, p. 25.

Dann, C. 1979. *The Animals of Farthing Wood*. London: Egmont Publishing.

Daston, L., and G. Mitman. 2005. *Thinking With Animals: New Perspectives On Anthropomorphism*. New York: Columbia University Press.

Daily Mash. 2012. 'Bring It, say Badgers', 18 September. Retrieved 24 May 2016 from http://www.thedailymash.co.uk/news-in-pictures/news-briefly/bring-it-say-badgers-2012091841472.

———. 2014 'Fudd Declares Badger Cull Vewy Successful', 6 January. Retrieved 24 May 2016 from http://www.thedailymash.co.uk/news/society/fudd-declares-badger-cull-vewy-successful-2014010682288.

Davies, G. 2000. 'Science, Observation and Entertainment: Competing Visions of Postwar British Natural History Television, 1946–1967', *Cultural Geographies* 7(4): 432–460.

DEFRA. 2011. 'Bovine Tuberculosis: The Government's Approach to Tackling the Disease and Consultation on a Badger Control Policy. Summary of Consultation Responses.' Retrieved 1 December 2014 from http://webarchive.nationalarchives.gov.uk/20120616115816/http://archive.defra.gov.uk/corporate/consult/tb-control-measures/bovinetb-summary-responses-110719.pdf.

———. 2014. 'A Strategy for Achieving Officially Bovine Tuberculosis Free Status for England.' Retrieved from https://www.gov.uk/government/ publications/a-strategy-for-achieving-officially-bovine-tuberculosis-free-status-for-england.

Donnelly, C.A., and R. Woodroffe. 2015. 'Bovine Tuberculosis: Badger-Cull Targets Unlikely to Reduce TB', *Nature* 526(7575): 640. doi:10.1038/526640c.

Douglas, M. 1966. *Purity and Danger: An Analysis of Concepts of Pollution and Taboo*. London: Routledge and Kegan Paul.

Ellis, A. 1877. 'The Badger' (Letters To The Editor), *The Times Archive*, issue 29081, 24 October, p. 5, col E.

Ellis, S. 2002. 'Tea for Two, Please', *The Times (Weekend)*, 6 July, p. 6.

Enticott, G. 2001. 'Calculating Nature: The Case of Badgers, Bovine Tuberculosis and Cattle', *Journal of Rural Studies* 17(2): 149–164.

———. 2008. 'The Spaces of Biosecurity: Prescribing and Negotiating Solutions to Bovine Tuberculosis', *Environment and Planning A* 40(7): 1568–1582.

Enticott, G. 2011a. 'The Local Universality of Veterinary Expertise and the Geography of Animal Disease', *Transactions of the Institute of British Geographers* 47(1): 75–88.

———. 2011b. 'Techniques of Neutralising Wildlife Crime in Rural England and Wales', *Journal of Rural Studies* 27(2): 200–208.

———. 2015. 'Public Attitudes to Badger Culling to Control Bovine Tuberculosis in Rural Wales', *European Journal of Wildlife Research* 61: 387–398. doi:10.1007/s10344-015-0905-9.

Farm Crisis Network. 2009. *Stress and Loss: A Report on the Impact of Bovine TB on Farming Families*. Farm Crisis Network, Northampton. Retrieved 1 December 2014 from http://www.tbfreeengland.co.uk/assets/4200.

Fissell, M. 1999. 'Imagining Vermin in Early Modern England', *History Workshop Journal* 47(Spring): 1–29.

Fitzgerald, S.D., and J.B. Kaneene. 2013. 'Wildlife Reservoirs of Bovine Tuberculosis Worldwide: Hosts, Pathology, Surveillance, and Control', *Veterinary Pathology* 50(3): 488–499.

Fricker, M. 2015. 'The culling fields; 2,000 badgers to be shot dead . . . & it will cost taxpayers £12million', *Daily Mirror*, September 25, p. 25.

Gerber D.L.J., C. Burton-Jeangros and A. Dubied. 2011. 'Animals in the Media: New Boundaries of Risk?' *Health, Risk & Society* 13(1): 17–30.

Goodwin, S. 1997. 'Cull or Cure Dilemma as Badgers get Blame for Epidemic', *The Independent*, 1 December, p. 20.

Grahame, K. 1908. *The Wind in the Willows*. London: Methuen.

Grahame, K., and P. Branscombe. 1913. *The Wind in the Willows*. New York: C. Scribner's Sons. Retrieved 1 December 2014 from *Project Gutenburg* at http://www.gutenberg.org/files/27805/27805-h/27805-h.htm.

Grahame, K., and I. Moore. 2009. *The Wind in the Willows (Walker Illustrated Classics)*. London: Walker Books.

Grahame, K., and E.H. Shepard. 1931. *The Wind in the Willows*. London: Methuen.

Grant, W. 2009. 'Intractable Policy Failure: The Case of Bovine TB and Badgers', *The British Journal of Politics & International Relations* 11(4): 557–573.

Griffin, E. 2007. *Blood Sport: Hunting in Britain since 1066*. New Haven, CT: Yale University Press.

Griffiths, H.I., and D.H. Thomas. 1997. 'The Conservation and Management of the European Badger (Meles meles)', *Nature and Environment, No. 90*. Strasbourg: Council of Europe Publishing.

Guardian Leader. 2007. 'Leading Article: In Praise of . . . Badgers', *The Guardian*, 15 May, p. 28.

Hall, M., and B.B. Cosgrove. 1984–1987. *The Wind in the Willows* (TV series). Manchester: Cosgrove Hall Productions.

Haraway, D. 2007. *When Species Meet*. Minneapolis: University of Minnesota Press.

Hardstaff, J.L., et al. 2014. 'Evaluating the Tuberculosis Hazard Posed to Cattle from Wildlife across Europe', *Research in Veterinary Science* 97(S): S86–S93.

Hardy, P. 1975. *A Lifetime of Badgers*. Newton Abbot: David and Charles.

Hattersley, R. 2005. 'A Cull by Royal Appointment', *The Times*, 22 December, p. 18.

Hill, C.M. 2015. 'Perspectives of "Conflict" at the Wildlife–Agriculture Boundary: 10 Years On', *Human Dimensions of Wildlife* 20(4): 1–6.

Hinsliff, G. 1998. 'Thousands of Badgers are Condemned to Die over TB Fears', *Daily Mail*, 18 August, p. 21.

Hytten, K. 2009. 'Dingo Dualisms: Exploring the Ambiguous Identity of Australian Dingoes', *Australian Zoologist* 35(1): 18–27.

Independent Scientific Group on Cattle TB. 2007. *Final Report of the Independent Scientific Group on Cattle TB*. London: Department of the Environment and Rural Affairs. Retrieved 1 December 2014 from http://archive.defra.gov.uk/foodfarm/farmanimal/diseases/atoz/tb/isg/report/final_report.pdf.
Jacques, B. 1988. *Mossflower*. London: Hutchinson.
Jenkins, S. 2004. 'A Verminous Vietcong Stalks the Countryside', *The Times*, 4 April, p. 24.
Joffe, H. 1999. *Risk and the Other*. Cambridge: Cambridge University Press.
Knight, J. (ed.). 2000a. *Natural Enemies: People–Wildlife Conflicts in Anthropological Perspective*. London: Routledge.
———. 2000b. 'Introduction', *Natural Enemies: People–Wildlife Conflicts in Anthropological Perspective*. London: Routledge, pp. 1–36.
Krebs, J.R., and the Independent Scientific Review Group. 1997. *Bovine Tuberculosis in Cattle and Badgers: Report to the Rt. Hon Dr. Jack Cunningham MP*. London: MAFF Publications.
Lascelles, E. 1932. 'The Badger: Damage Caused in the Hunting Countries', letter from Edward Lascelles, *The Times Archive*, issue 46226, 31 August, p. 6, col. B.
Law, J., and M. Lynch. 1988. 'Lists, Field Guides, and the Descriptive Organization of Seeing: Birdwatching as an Exemplary Observational Activity', *Human Studies* 11(2): 271–303.
Leach, M., I. Scoones and A. Stirling. 2010. 'Governing Epidemics in an Age of Complexity: Narratives, Politics and Pathways to Sustainability', *Global Environmental Change* 20(3): 369–377.
Lester, L. 2010. *Media and Environment: Conflict, Politics and the News*. Cambridge: Polity.
Lewis, C.S. 1951. *Prince Caspian: The Return to Narnia*. London: Geoffrey Bles.
Lodge, M., and K. Matus. 2014. 'Science, Badgers, Politics: Advocacy Coalitions and Policy Change in Bovine Tuberculosis Policy in Britain', *Policy Studies Journal* 42(3): 367–390.
Lorimer, J. 2007. 'Nonhuman Charisma', *Environment and Planning D: Society and Space* 25: 911–932.
Lovegrove, R. 2007. *Silent Fields: The Long Decline of a Nation's Wildlife*. Oxford and New York: Oxford University Press.
Marcu, A., E. Lyons and P. Hegarty. 2007. 'Dilemmatic Human-Animal Boundaries in Britain and Romania: Post-Materialist and Materialist Dehumanization', *British Journal of Social Psychology* 46(4): 875–893.
Marren, P. 1995. *The New Naturalists: Half a Century of British Natural History*. London: HarperCollins.
Marvin, G. 2000. 'The Problem of Foxes: Legitimate and Illegitimate Killing in the English Countryside', in J. Knight (ed.), *Natural Enemies: People–Wildlife Conflicts in Anthropological Perspective*. London: Routledge, pp. 189–211.
Marvin G. 2001. 'Cultured Killers: Creating and Representing Foxhounds', *Society and Animals* 9: 273–292.
Mavhunga, C.C. 2011. 'Vermin Beings: On Pestiferous Animals and Human Game', *Social Text* 29(1): 151–176.
May, B. 2010. *Save Me* (campaign website). Retrieved 1 December 2014 from www.save-me.org.uk.
Maye, D., et al. 2014. 'Animal Disease and Narratives of Nature: Farmers' Reactions to the Neoliberal Governance of Bovine Tuberculosis', *Journal of Rural Studies* 32: 401–410. doi:10.1016/j.jrurstud.2014.07.001.
McDonald, R.A. 2014. 'Animal Health: How to Control Bovine Tuberculosis', *Nature* 511(7508): 158–159. doi:10.1038/nature13514.

Mitchell, S. 2006. 'TB OR NOT TB? Badgers are on Trial, Accused of Spreading Tuberculosis. Before the Jury is Dismissed on Friday, Sandy Mitchell Talks to both Defence and Prosecution', *Daily Telegraph(Weekend)*, 4 March, p. 1.

Molloy, C. 2011. *Popular Media and Animals*. Basingstoke: Palgrave Macmillan.

More, S.J. 2009. 'What is Needed to Eradicate Bovine Tuberculosis Successfully: An Irish Perspective', *Veterinary Journal* 180(3): 275–278.

Mort, M., et al. 2008. 'Animal Disease and Human Trauma: The Psychosocial Implications of the 2001 UK Foot and Mouth Disease Disaster', *Journal of Applied Animal Welfare Science* 11(2): 133–148.

Moss, S. 2004. *A Bird in the Bush: A Social History of Birdwatching*. London: Aurum.

Muirhead, R.H. 1972. 'Bovine Tuberculosis in Wild Badgers in South Gloucestershire', *State Veterinary Journal* 27(1), September: 197–205.

Muirhead, R.H., J. Gallagher and K.J. Burn. 1974. 'Tuberculosis in Wild Badgers in Gloucestershire: Epidemiology', *Veterinary Record* 95(24): 552–555.

Müller, B. et al. 2013. 'Zoonotic Mycobacterium Bovis-Induced Tuberculosis in Humans', *Emerg Infect Dis* 19(6): 899–908.

Naylor, R., et al. 2015. 'The Framing of Public Knowledge Controversies in the Media: A Comparative Analysis of the Portrayal of Badger Vaccination in the English National, Regional and Farming Press', *Sociologia Ruralis* Early View. doi:10.1111/soru.12105.

Neal, E. 1948. *The Badger (New Naturalist Monograph)*. London: Collins.

———. 1994. *The Badger Man: Memoirs of a Biologist*. Ely: Providence Press.

Neal, E., and H. Hewer (filmmakers). 1954. *Badgers*. BBC, British Broadcasting Corporation. Retrieved from http://genome.ch.bbc.co.uk/c28140ea6560458696f962560142258c.

Nelson, M. 1975. 'Old English Riddle no. 15: "The Badger", an Early Example of Mock Heroic', *Neophilologus* 59(3): 447–450.

Nerlich, B., B. Brown and N. Wright. 2009. 'The Ins and Outs of Biosecurity: Bird Flu in East Anglia and the Spatial Representation of Risk', *Sociologia Ruralis* 49(4): 344–359.

Nerlich, B., and R. James. 2009. '"The Post-Antibiotic Apocalypse" and the "War on Superbugs": Catastrophe Discourse in Microbiology, its Rhetorical Form and Political Function', *Public Understanding of Science* 18: 574–590.

Nicholls, H. 2011. 'The Art of Conservation', *Nature* 472(7343): 287–289.

O'Connor, C.M., D.T. Haydon and R.R. Kao. 2012. 'An Ecological and Comparative Perspective on the Control of Bovine Tuberculosis in Great Britain and the Republic of Ireland', *Preventive Veterinary Medicine* 104(3–4): 185–197.

Pease, A.E. 1898. *The Badger: A Monograph*. London: Lawrence and Bullen Limited.

Perkins, A. 1999. 'Inside Story: What Have I Done to Deserve This? The Government is About to Launch a Badger-Culling Campaign. But Instead of Solving a Problem, it's Likely to Cause a New Catastrophe in the Countryside, says Anne Perkins', *The Guardian*, 12 April, p. 8.

Perkins, D. 2003. *Romanticism and Animal Rights (Cambridge Studies in Romanticism)*. Cambridge and New York: Cambridge University Press.

Picking, J. 2003. *Badger Badger Badger* (web animation). Retrieved 1 December 2014 from http://weebls-stuff.com/songs/badgers/.

Picking, J., B. May and B. Blessed. 2013. *Save The Badgers* (web animation). Retrieved 1 December 2014 from http://www.weebls-stuff.com/songs/Save+The+Badgers/.

Political Scrapbook. 2013. 'Best of the Badger: The 5 Funniest Responses to Paterson's Own Goal', *Political Scrapbook*. Retrieved 1 December 2014 from http://politicalscrapbook.net/2013/10/5-funniest-responses-to-badger-goalposts-gaffe/.

Potts, A. 2009. 'Kiwis Against Possums: A Critical Analysis of Anti-Possum Rhetoric in Aotearoa New Zealand', *Society and Animals* 17: 1–20.

Potter, B. 1912. 'The Tale of Mr. Tod.' London: Frederick Warne & Co. Retrieved from Project Gutenburg at http://www.gutenberg.org/files/19805/19805-h/19805-h.htm.

Potter, B., and J. Taylor. 2012. *Beatrix Potter's Letters* (ebook edition). London: Penguin Books Limited. Retrieved 1 December 2014 from http://books.google.com/books?id=BvMuYoVIZugC&pgis=1.

Redpath, S.M., S. Bhatia and J. Young. 2014. 'Tilting at Wildlife: Reconsidering Human–Wildlife Conflict', *Oryx* 49(2): 222–225. doi:http://dx.doi.org/10.1017/S0030605314000799.

Rees, A. 2006. 'A Place that Answers Questions: Primatological Field Sites and the Making of Authentic Observations', *Studies in the History and Philosophy of Science* C 37: 311–333.

Robinson, P.A. 2015. 'A History of Bovine Tuberculosis Eradication Policy in Northern Ireland', *Epidemiology and Infection* 143(15): 3182–3195. doi:10.1017/S0950268815000291.

Roper, T.J. 2010. *Badger* (The New Naturalist Library). London: Collins.

Rowling, J.K. 1997. *Harry Potter and the Philosopher's Stone*. London: Bloomsbury.

Runhaar, H., M. Runhaar and H. Vink. 2015. 'Reports on Badgers *Meles meles* in Dutch Newspapers 1900–2013: Same Animals, Different Framings?' *Mammal Review* 45(3): 133–145.

Schlich, T., E. Mykhalovskiy and M. Rock. 2009. 'Animals in Surgery – Surgery in Animals: Nature and Culture in Animal–Human Relations and Modern Surgery', *History and Philosophy of the Life Sciences* 31(3–4): 321–354.

Scoones, T. 2009. *Springwatch, Episode 2, Predatory Badgers* (TV series). Bristol: BBC Natural History Unit. Retrieved 1 December 2014 from http://www.bbc.co.uk/nature/life/European_Badger#p007x3mn.

Spencer, A. 2011. 'One Body of Evidence, Three Different Policies: Bovine Tuberculosis Policy in Britain', *Politics* 31(2): 91–99.

Spencer, M. 2010. 'Imagining Badgers: An Attempt at Working with Objects of Governance', *Sentient Creatures: Transforming Biopolitics and Life Matters Conference Oslo*, 16–17 September. Oslo: University of Oslo.

Stirling, A. 2008. 'Opening Up and Closing Down: Power, Participation and Pluralism in the Social Appraisal of Technology', *Science Technology and Human Values* 33(2): 262–294.

Strauss, A., and J. Corbin. 1998. *Basics of Qualitative Research: Techniques and Procedures for Developing Grounded Theory*, 2nd edition. Thousand Oaks, CA, and London: Sage Publications.

Syse, K.V.L. 2013. 'Otters as Symbols in the British Environmental Discourse', *Landscape Research* 38(4): 540–552.

Tasker, J. 2012. 'In the Hot Seat: Owen Paterson', *Farmer's Weekly Interactive*, 14 September. Retrieved 1 December 2014 from http://www.fwi.co.uk/articles/14/09/2012/135217/in-the-hot-seat-owen-paterson.htm.

The Times. 1811. 'Easter Monday Sports', *The Times Archive*, issue 8271, 16 April, p. 3, col C.

———. 1927. 'Men And Badgers (Editorial)', *The Times Archive*, issue 44567, 28 April, p. 15, col E.

———. 1999. 'Badger the Government', *The Times*, 5 May.

Thomas, E. 1917. *Poems*. London: Selwyn & Blount.

Waddington, K. 2006. *The Bovine Scourge: Meat, Tuberculosis and Public Health*. Woodbridge: The Boydell Press.

Washer, P. 2010. *Emerging Infectious Diseases and Society.* Basingstoke: Palgrave Macmillan.
Weathers, H. 2003. 'Mother Badgers are Snared in Rusty Cages, Parted from their Screaming Cubs and Coldly Shot in the Head . . . all with the Government's Blessing', *Daily Mail*, 3 June, p. 11.
White, T.H. 1938. *The Sword in the Stone.* New York: G.P. Putnam's Sons.
White, P.C.L., and A.I. Ward. 2010. 'Interdisciplinary Approaches for the Management of Existing and Emerging Human–Wildlife Conflicts', *Wildlife Research* 37(8): 623.
Wilkinson, K. 2007. 'Evidence Based Policy and the Politics of Expertise: A Case Study of Bovine TB', *CRE Discussion Paper* 12 (Working paper). Retrieved 1 December 2014 from http://www.ncl.ac.uk/cre/publish/discussionpapers/pdfs/dp12%20Wilkinson.pdf.
Woodroffe, R., et al. 2009. 'Social Group Size Affects Mycobacterium Bovis Infection in European Badgers (*Meles meles*)', *Journal of Animal Ecology* 78: 818–827.
Woods, M. 2000. 'Fantastic Mr. Fox? Representing Animals in the Hunting Debate', in C. Philo and C. Wilbert (eds), *Animal Spaces, Beastly Places: New Geographies of Human–Animal Interactions.* London: Routledge, pp. 183–202.
———. 2008. 'Hunting: New Labour Success or New Labour Failure?' in M. Woods (ed.), *New Labour's Countryside: Rural Policy in Britain since 1997.* Birmingham: The Policy Press, pp. 95–114.
Veterinary Association for Wildlife Management. 2010. *Badgers and Bovine TB.* Retrieved 26 November 2014 from http://www.vet-wildlifemanagement.org.uk/index.php?option=com_content&task=view&id=14&Itemid=28.

5
Savage Values
Conservation and Personhood in Southern Suriname

Marc Brightman

In their introduction to the 2005 Darwin College Lectures under the theme of conflict, Martin Jones and Andrew Fabian contrast harmony and equilibrium on one hand, and 'disharmony' and 'conflict' on the other. As they point out, 'disharmony and conflict may be far more than aberrations from a normal state; they may be at the heart of the system' (2006: 1). One might assume that an ecological system will not survive for long if it is in disequilibrium, and if conflict is at its heart. The idea of indigenous peoples of Amazonia living in harmony with their environment may be a cliché, with origins in romantic portrayals of native Americans such as those of Hudson, Rousseau or Chateaubriand, but recent studies in anthropology and ecology have confirmed that indigenous Amazonian ways of life are materially 'sustainable', given the right circumstances (Shepard et al. 2012), although they are not necessarily maintained on the basis of ideologies of sustainability. Nor are they based on ideas of harmony or equilibrium; indeed, students of Amerindian thought have found it characterized on the contrary by a tendency towards 'perpetual disequilibrium' (Lévi-Strauss 1991: 316). Change, but also social reproduction, may be regarded as being due to the fact that the elements of the whole are in disequilibrium, even when their immediate relationships towards each other appear stable. In this chapter I shall try to explore the implications of these features of native Amazonian society and ecology for ideas about human-wildlife conflict. I shall begin with some reflections about the distinction between nature and culture heavily implicit in the 'human-wildlife' relationship. I shall then consider what it means to be 'human' and to be 'animal' in native Amazonian societies. This will bring me to the specific case of the Trio, who, as I shall describe, have become involved first with Christianity and then with conservation organizations. I shall argue that the notion of human-wildlife conflict is implicit in the conservation agenda, but depends on a worldview that derives from Christianity. Yet, as I will show, the Trio's engagement with conservation and the human-wildlife conflict paradigm is based on their own agenda, not one prescribed by outsiders, as I shall briefly discuss by way of conclusion.

It is now well established among scholars of Amazonian historical ecology that the native peoples of the region manage its natural resources rather than merely exploiting them, making their environment a cultural landscape. This contrasts starkly with the long-held assumption that indigenous peoples merely 'adapted' to their natural environment in a haphazard way adequate only for bare subsistence. The management of game species is of great importance for native Amazonians, for whom the hunting and eating of game play important cultural and nutritional roles. Hunting in and near swiddens (forest gardens) has occurred since pre-Colombian times, some groups burn grassland to stimulate new growth to attract game and many Amazonian groups 'explicitly recognize that swiddens, swidden/forest ecotones, and other manipulated zones lure important game species'. Peccaries are one example of a game animal that has been so carefully managed by native people that they have been described as 'semi-domesticated' (Balée 1989: 15).[1] This use of swiddens to attract game is explicitly referred to by native peoples of Amazonia, who deliberately overplant to compensate for losses to animals (Carneiro 1983; Posey 1984; Balée and Gely 1989; Dufour 1990; Shepard et al. 2012: 654; Zent 2012).

This widespread cultural practice of welcoming non-human animal species to gardens, even when they are attracted to the same food plants as humans themselves, contrasts starkly with the typical scenario of human-wildlife conflict, in which wild animals are regarded as competing against humans for the same resources, whether they be carnivores that threaten domesticated animals or herbivores such as rabbits that deplete food crops. This difference has a certain amount to do with the level of pressure on space. Economic development, industrialization, land use intensification and frontier expansion lead to the elimination or control of animal populations, and all over the world '"explorers" and prospectors led the way for commodity extractors; settlers ploughed grasslands or semi-arid lands, drained wetlands, built dams, and cleared forests' (Emel and Wolch 1998: 4). However, it also has much to do with the ways in which humans engage with nature. As Kay Milton has argued, non-utilitarian conservation, or 'preservationism', which seeks to protect nature as an end in itself, can be understood as a kind of symbolic practice, based on a dualistic view of nature and culture. Its purpose is to try to protect the boundary between nature and culture (Milton 2000, in Knight 2003: 17).

Milton's suggestion represents one useful way of thinking about human-wildlife conflict in terms of cultural categories. It can be read in counterpoint with critiques of prevailing assumptions about the separateness of nature and culture based on ethnographies of native Amazonian swidden horticulturalists (Descola 1996) or arctic hunter gatherers (Ingold 2000). Out of this critical literature on nature and culture, there has emerged an approach to the anthropology of non-humans that treats the latter as social actors in their own right – a new tendency to treat 'animals as *parts* of human society rather than just *symbols* of it' (Knight 2005, in Candea 2010: 243; cf. Kohn 2013). This point should be understood in the context of the now widespread consensus across the social sciences that the nature-society or nature-culture dichotomy has been socially and historically constructed. Historians and philosophers of science have shown how an objective 'nature' is constructed in laboratories. This process of isolation and reduction is highly effective at producing useful knowledge, but it ironically leads to the emergence of more and more nature-culture 'hybrids', as Bruno Latour calls them, from which further 'natures' must in turn be precipitated (1991). The Cartesian separation of nature and culture, so characteristic of the modern age, has been projected upon landscapes in the form of 'land-sparing' strategies that allocate spaces for human and non-human activity.[2] In this light, nature reserves, far from being spaces for natural ecological successions to take place independently of human activity, are quintessential expressions of the Anthropocene; indeed, they are examples of modernity writ large. More concretely, in practice, the attempts to construct boundaries between 'wild' natural and human ecologies frequently relies on coercion, which can lead to human conflict (Adams, forthcoming; Homewood, forthcoming) and can even be counterproductive in terms of biodiversity conservation as they intensify pressure upon traditional grazing and agricultural land.

The Human-Wildlife Conflict Collaboration notes that 'human-wildlife conflict is as much a conflict between humans and wildlife, as it is a conflict *between* humans *about* wildlife' (HWCC 2012, original emphasis). The phrasing of this quotation is worth examining more closely. It implies that the fact that human-wildlife conflict is a conflict 'between humans and wildlife' is taken as given – what the reader is being told is that, in addition to this, it is also a conflict '*between* humans *about* wildlife'. This makes it appear as though the opposite shift has taken place in the area of human-wildlife conflict practitioners from that which has occurred in anthropology. If anthropologists have recently learned to begin thinking about animals as actors in their own right, entering into direct relationships with humans (whether conflictual or not), conservationists in the area of human-wildlife conflict would appear to have started with the assumption that the primary relationship is between humans and animals, and to have then learned to think about relationships between humans with respect to animals. However, this interpretation masks the fact that the only kind of relationship between humans and animals being considered here is one of conflict. This brings us back to the accusation against preservationism that it serves merely to reinforce artificial, culturally situated boundaries between nature and culture. But as Catherine Hill, Ferrel Osborn and Andrew Plumptre have noted, 'there is a very real need to consider human-wildlife conflict issues within the context of local community and individual needs as well as conservation objectives' (2002: 11). The local context and the intersections of individual needs may often muddy the waters and blur presumed boundaries between nature and culture.

Personhood in Amazonia

There is now a large amount of ethnographic evidence that shows that human personhood is by no means regarded by indigenous Amazonian people as 'given', or dictated by the facts of birth and biological kinship. Babies are in a fragile ontological state, and risk being transformed into non-humans at any moment; their bodies must be moulded into real human bodies, sometimes literally by physical manipulation, as though they were pieces of clay that had to be shaped into human form (Lagrou 2000; McCallum 2001; Vilaça 2005). Throughout life, it is the daily interaction between co-residents that maintains their shared humanity – especially eating together – and indeed commensality not only indicates but also produces shared identity (Vilaça 2002).

This fragile, processual and constructed nature of human personhood is grounded in a native ontology widely reported across Amazonia, according to which humanity is a condition that is subjectively shared by all species, while animality is a condition of alterity. Because human-animal relations are thus defined by the subject's point of view, this has come to be known as Amerindian perspectivism (Viveiros de Castro 1998). It has a strong association with hunting and, more broadly, with predation, which can be expressed thus: those who eat humans, and who are eaten by humans, instead of eating *with* them, are animals (Fausto 2007). Animals, however, also share humanity with each other, from their own point of view. Shamans frequently encounter animals not as beasts but as other persons, and in order to do so they must compromise a part of their own humanity, or rather, a part of their own human perspective. The boundaries between humanity and animality are blurred through such encounters, and this is exposed through the evidence of shared social orders, in the form of practices and structures of kinship.

When the logic of being is such that humanity is a relative concept, there is a need to emphasize that one belongs to the category of real human beings, if only as a further expression of the need constantly and actively to maintain human status. It is perhaps for this reason that many ethnonyms across the region have markers emphasizing that they are not just people, but *real* people or *true* people (Clastres 2010), or else the inclusive pronoun 'we', implying an opposition to the category of animal, prey, food and enemies (Vilaça 2002: 354). However, encounters with animality are necessary for the production of humanity, and as Aparecida Vilaça has written, 'humanity is conceived of as a position, essentially transitory, which is continuously produced out of a wide universe of subjectivities that includes animals' (2002: 349).

In many, perhaps most, native Amazonian cosmologies, certain non-human species – particularly game animals – are owned by spirit-masters who, in many cases, negotiate the supply of game with human shamans (Århem 1996). As Carlos Fausto has written:

> This category of owner-master is widespread in the region and corresponds to what Hultkrantz (1961) termed 'the supernatural owners of nature'. Until recently, ethnology limited itself to these figures when speaking of owners or masters, depicting them as hyperboles of the species they represent or the anthropomorphic form through which they appear to shamans. These figures need to be reinserted in the overall set of ownership relations, since, as Cesarino notes in relation to another Panoan people, the masters of animals 'replicate the same configuration that characterizes the

Marubo maloca owners (*shovō ivo*): both are chiefs of their houses, in which they live with their families and have their own ways of being' (Cesarino 2008:25). The animal masters are owners in their own environment, containing a species-collectivity within themselves. (Fausto 2008: 332–333).

Fausto points out that 'the owner-master is *the form through which a plurality appears as a singularity to others*' (emphasis added), whether it is the master of animals and his 'children', or a shaman and his spirit familiars, or a village leader and his followers (Fausto 2008: 335).

Philippe Descola has led the way in recent decades in reviving the concept of animism as a way of understanding a widespread human relationship with 'nature' characterized by the belief that non-human species live in human-like 'societies'. Descola points out that Amazonian 'animic systems are closed systems, within which circulate all the elements necessary for the maintenance of the cosmos' (Rivière 2001: 33, my translation). There has been some discussion of the fact that, within Amazonia, groups which in material terms relate to the living environment in more or less identical ways, may conceive of their practices very differently. Descola contrasts two modes of relating to non-human others, especially prey animals – predatory (for which he uses his own ethnography of the Jívaro) and reciprocal (for which he takes the example of the Tukanoan-speaking Desana, using the ethnography of Reichel-Dolmatoff). Desana shamans have the role of maintaining the equilibrium of souls by negotiating with masters of animals – who appropriate the souls of human dead and keep these souls in the aquatic underworld. Humans are thus exchanged for animals (Rivière 2001). Jívaroans conceive of the human realm as a closed system, and used to respond to the death of their own by headhunting – they must capture human substance and life force by killing people who are not in their own group, but not too alien either – i.e. other Jívaroan groups. Even though ensuing deaths would tend to lead to reciprocal headhunting expeditions, killing was conceived of as a one-sided act, an act of predation and not an act of reciprocity. Hunting is understood in the same way by Jívaroans – as an act of predation, without any need for reciprocity (Rivière 2001).

However, the difference between these types proposed by Descola is less marked in reality. Rivière points to Århem's work on the Makuna, another Tukanoan group, who, he says, conceive of their relations with animals in terms of predation: the cosmos is conceived of as a hierarchy of beings with jaguar and anaconda gods at its apex. These bring about death and reproduction in humans, and by the same logic human predation of animals brings about the possibility of the renewal and reproduction of the category to which it belongs. In the Guiana region, for instance among the Trio, there is no evidence for the exchange of human for animal souls – of reciprocity – instead, the circulation of souls through death and birth to and from the reserve of souls on the eastern horizon appears to occur by itself, without intervention from third parties (Rivière 2001: 40). Nevertheless, the *cause* of death tends to be witchcraft, or soul theft, that is, action by spirit or human enemies from outside – by definition, for only outsiders practice spirit attacks. In other words, the cause of death is predatory action. Meanwhile reciprocity has a role in renewal of society through ritual action.

Peter Rivière (2001) illustrates these differences through three myths concerning Trio relations with prey animals. Two similar myths, in which a shaman meets a tapir and a spider monkey respectively, involve poor communication with the animals in question – they meet by accident; the animals put the shaman in a hammock with his own daughter but forbid him from touching her; the tapir/spider monkey gives the shaman a great deal of information but forbids him from telling anything of his encounter to his kin – but the information comes out eventually when the shaman's tongue is loosened with drink – the villagers look for the tapir/monkey; and the shaman protagonist pays with his life. In another myth, a shaman goes to look for peccaries, this time on purpose. He meets the master of peccary, and they address each other as trading partners (emphatically not affines, unlike the other cases that involve quasi fathers-in-law), and they come to an arrangement whereby the peccaries will be released for people to hunt.

Trio see peccaries as social animals, much more social – and hence human-like – than spider monkeys or tapirs. A further difference is that there is no indication that the Trio's relationship with the animals in the first myth is with masters of animals; instead it is with the animals themselves. In the case of the peccaries, in contrast, there are all the signs of a shamanic journey – a path opens into the underworld, and there is a reference to the master of peccaries. Rivière points out that relations with non-humans for the Trio can thus take differing forms: predatory or reciprocal, according to each case, and that in this they reflect Trio social relations between neighbouring settlements, which range from friendly to hostile. Among other Amazonian cases, there are similar variations.

During my own recent fieldwork among the Trio, I found that their attitudes towards hunting are now even more ambiguous. Most Trio do not see any cosmological reason for not hunting as much as they like. However, there is a strong stigma attached to buying and selling meat – which is unsurprising when one considers that to buy and sell meat would efface the element of sharing, of commensality, and hence of shared identity. A clear confirmation of this can be seen where meat is indeed bought and sold, in some Wayana villages on the border with French Guiana. Here, such trade only occurs between individuals without kin ties, when Wayana traders go to distant villages carrying the meat in refrigerators balancing in their dugout canoes. Hunting and spirit attacks also seem to be linked: men who overhunt are said to be punished with disease, and two men in the extended household that hosted me during my fieldwork were said to have lost their wives as a result of overhunting: their wives died of cervical cancer, but the disease resulted from a spirit master of animals taking revenge for their greed in killing too many of his children.

In a sharing economy, there would of course be no need to overhunt, unless population pressure grew very high, but this is very far from being the case in southern Suriname. However, the concentration of Trio people in large villages around an airstrip, health clinic, church and school over the last half century has led to many changes. One of these is access to markets, and there is a thriving market for bush meat. This is sent regularly to the capital city, Paramaribo, by aeroplane, and since most aircraft come bringing manufactured objects to the village there is always space, at a discounted rate, for sending such high-value forest products to the city.

Changing Cosmologies

Something that undoubtedly contributes to the Trio's willingness to overhunt is the new cosmology that has been offered to them through conversion to Christianity. The gulf between the Judeo-Christian worldview and that of native Amazonians is wide, but certain concepts do seem to have taken hold. One of these is that Jesus is a kind of supreme master-owner of all creatures – all people are his children, and hence all men are brothers. Meanwhile all animals are under his dominion. The notion that men can therefore hunt unlimited quantities of prey may not automatically follow; however, some missionaries have done their best to encourage such an idea. One made a point of shooting all the caimans he could find, knowing that the caiman is an important mythical figure for the Trio. He came back with large quantities of meat and, since at the time there were no regular flights to the city allowing a bushmeat trade to thrive, much of it rotted. Another missionary, still active today, who prefers to be known as a schoolteacher, told me how he would try to bring Trio to camp in places supposed to be the 'houses' or 'villages' of spirit-master/owners, to show through rational experiment that these sacred places could be visited without spiritual danger.

In the world of conservation, there is an important field of imagery and rhetoric portraying native Amazonians as guardians of the forest – as 'ecological Indians'.[3] At the same time, some native Amazonian groups, or at least certain native leaders, have learned to exploit the image of the ecological Indian to attract funding and 'projects' that bring employment and access to manufactured objects. In other words, they have moved outside any pre-existing cultural propensity to live in harmony with nature that they may have had, to exploit global and national images of themselves as guardians of biodiversity. In reality this is less of a calculated move than it may appear, and over time it can be understood as a cultural shift. Certain commentators have pointed to the 'middle ground' on which native leaders and environmentalists can communicate to achieve often divergent goals. The fault lines emerging in this middle ground are related to the heavy semiotic load that Amazonian Indians carry as actors on the global political stage, representing core values of the imagined eco-community. Representations of Amazonian Indians circulating in the international public sphere tend to be generic stereotypes that misrepresent the diversity of native Amazonian cultures and the complexity of native priorities and leadership issues. More importantly, generic representations, no matter how sympathetic, inevitably turn into liabilities when the disjunctures between external images and indigenous realities become manifest (Conklin and Graham 1995).

The reality is complex, and I do not want to argue that native Amazonians do not care for their environments or manage them in sustainable ways – in fact, much evidence, comparing indigenous reserves with protected areas, suggests that forest peoples are more successful at maintaining biodiversity and forest cover than conservation professionals (Nelson and Chomitz 2011). However, it is clear that, in so far as they do, they do so for different reasons than environmentalists.

To illustrate this: in Kwamalasamutu, the largest Trio village, the population had reached over 1,500 – about fifty times the size of traditional settlements. According to most accounts, it became clear that the degradation of the surrounding forest was becoming too serious to ignore. Hunters had to travel with their families on expeditions lasting as long as two weeks to indulge in what a researcher from Paramaribo described to me as 'protein binges', because of the distance that had to be travelled to achieve reasonable rates of hunting success.[4] The village leader began to discuss the matter with the head of a major international conservation organization, which had been active in the region for a long time but which has only relatively recently begun to engage in cooperation with indigenous peoples. He asked her if her organization could lend its expertise to help him to create a farm. White people manage to have plenty of meat because they domesticate their animals; they farm them and produce as much as they need. Why not do the same with peccary, agouti and spider monkeys?

The head of the organization told me that she 'explained' that domesticating these animals is not so simple. She may also have known that missionaries had tried to introduce cows and goats among the Trio, without success (except on the savannah across the border in Brazil), but 'of course', as she said, turning a Trio village into a ranch would be the last thing she would want to do. Instead, she took the opportunity to suggest creating a restricted area to be managed by the Trio themselves – to experiment with a designated area for game animals to renew their populations, where hunting would be banned. This now appears to be having some success.

However, it also illustrates the gulf between different perspectives on nature. The Trio are struggling to adapt to a situation in which they live in much more concentrated centres of population, and hunt more than they used to in order to sell meat for cash. There are two principal causes of this situation, which are both aspects of the creation of mission stations: the creation of the mission stations in the first place, in the 1960s, was motivated by a desire on the part of the state to control the border areas of the interior and to 'civilize' their native inhabitants, but was never informed by considerations of the ecological consequences of moving small-scale, semi-nomadic swidden horticulturalists and hunter gatherers into large villages. The evangelization of these groups has also led, at best, to confusion about the order of things in the cosmos.

Conclusion

Today, conservation is increasingly becoming linked to questions of land rights and traditional livelihoods, through the internationalization of conservation and the 'green economy', most notably through the development of the REDD+ (Reducing Emissions from Deforestation and Forest Degradation) mechanism, which would seek to reduce deforestation by linking forests to global carbon markets.[5] Suriname is currently working hard, with the help of green NGOs such as Conservation International (CI), to meet the requirements of the UN and World Bank to become eligible for REDD+ 'readiness' funding. This has direct consequences for biodiversity conservation, which is included as part of the 'safeguards' element of REDD+; indeed, the degradation of forests is one of the things that REDD+ is designed to reduce. This brings increasing requirements for checks and balances, most notably the requirement for Free and Prior Informed Consent (FPIC), which raises some interesting questions.

FPIC makes an appeal to the 'middle ground', but does more than this, because, to be carried out properly it needs to involve informing and obtaining the consent of the entire community. However, communication on this level can scarcely happen in a meaningful way unless all parties share worldviews, or at least have overlapping worldviews. There is an important sense, therefore, in which the 'ecological Indian' who consents to green development must be a Christian – or must at least partially share a Euro-American model of nature and culture, whereby cultures are multiple and nature is universal, all men being brothers by virtue of their shared descent from Adam or ape.

Ironically, it is also this very same adaptation to a Euro-American world view that legitimizes overhunting and engagement in activities such as the illegal trade in exotic animals. So I would suggest that the ecological and irresponsible 'Indian' are two sides of the same coin, which is the struggle of native peoples to adapt to modernity. Human-wildlife conflict, then, is a 'modern' problem, or a problem of our times. It is an idea that emerges from the tendency to measure and calculate, characteristic of modernity. But, more than this, it emerges from the tendency to demand close monitoring of resources on multiple scales, to model risk, to put in place accounting and auditing mechanisms. 'Conflict' here means an ecological imbalance and a conflict of interests, identified through scientific data and equilibrium models. It is only possible to see deforestation and forest degradation as a problem in Suriname – a country with 90 per cent forest cover – if one is aware of trends in global deforestation and scientific estimates of their consequences.

The Trio have their own solution to local forest degradation: small units of nuclear and extended families have begun to leave large villages and create new ones in areas where resources are not depleted – and which tend to be the locations of former villages that existed before the process of sedentarization began. Human-wildlife conflict does not exist for them; but rather, the need to continue to reaffirm their shared humanity by living well together, killing prey and eating and drinking with their kin. The idea of human-wildlife conflict relies on a distinction between nature and culture that is far from universal. Indeed, the assumption of such a dichotomy may be part of the cause of the conflict between the two that is so widely observed. In contrast, animist peoples such as the Trio are perhaps able to 'manage' game and other resources with greater success precisely because they regard game animals as distant others sharing the same sensibilities as themselves. But it is worth remembering that the Trio are privileged to have the option of reacting to environmental degradation and political conflict alike (which for them are not such different things) by relocating. Indeed, part of the reason for the repopulation of old villages dispersed around Trio territory is to lay claim to land rights, which are under negotiation at present with the government. Suriname is the last country in South America to remain without any official recognition of indigenous land rights. This underlines the fact that the real source of conflict over resources in Suriname and elsewhere is not something basic to the difference between humans and nature. Its origin is in the conflict of interests between the powerful and the powerless among humans.

Marc Brightman is lecturer in Social Anthropology and co-director of CAOS, the Centre for the Anthropology, of Sustainability, at University College London. His research interests include global political ecology, farming and migration. His books include *The Imbalance of Power: Leadership, Masculinity and Wealth in Amazonia* and the edited volumes *Animism in Rainforest and Tundra*, with Vanessa Grotti and Olga Ulturgasheva, and *Ownership and Nurture: Studies in Native Amazonian Property Relations*, with Carlos Fausto and Vanessa Grotti.

Notes

1. Balée notes: 'Small and dispersed swiddens probably produced ecotonal environments which favoured higher densities of peccaries in addition to other important game species, such as tapir, deer, paca, and agouti, than would otherwise have been so. Semi-domesticated animals were lured not only to domesticated crops, but also to semi-domesticated (and sometimes deliberately planted) tree species. Indigenous peoples often plant such species with the declared intention of baiting game animals' (1989: 16).
2. See E.O. Wilson's recent book (2016) for an especially audacious call for land-sparing on a grand scale.
3. See Redford (1991) for a critique of romanticized notions about indigenous resource use.
4. One should not necessarily assume that this depletion of resources is simply due to population increase. Changes in hunting technology (shotguns rather than bows and arrows) in fact have a greater overall impact on game resources. Moreover, 'distribution of human settlements in the landscape is more important than population growth per se' (Shepard et al. 2012: 658), and indeed this seems to be the key factor in this case.
5. There is a vast amount of academic and grey literature on this subject. For official information, see www.unredd.org. For access to numerous resources as well as a critical introduction to REDD, see www.reddmonitor.org. The question of how the mitigation of forest destruction and degradation would be financed under REDD+ is highly contentious and many argue that it should not be done through market mechanisms. However, the concept of REDD has from the outset been based on a market logic that forests should be protected because of their 'true' value as natural capital.

References

Adams, W. Forthcoming. 'Conservation from Above: Globalising Care for Nature', in M. Brightman and J. Lewis (eds), *Beyond Development and Progress: Anthropological Visions of Sustainable Futures*. New York: Palgrave.

Århem, K. 1996. 'The Cosmic Food Web: Human-Nature Relatedness in the Northwest Amazon', in P. Descola and G. Pálsson (eds), *Nature and Society: Anthropological Perspectives*. London: Routledge, pp. 185–204.

Balée, W. 1989. 'The Culture of Amazonian Forests', in D. Posey and W. Balée (eds), *Resource Management in Amazonia: Indigenous and Folk Strategies*, a special edition of *Advances in Economic Botany* 7: 1–21.

Balée, W., and A. Gély. 1989. 'Managed Forest Succession in Amazonia: The Ka'apor Case', in D. Posey and W. Balée (eds), *Resource Management in Amazonia: Indigenous and Folk Strategies*, a special edition of *Advances in Economic Botany* 7: 129–158.

Candea, M. 2010. 'I Fell in Love with Carlos the Meerkat: Engagement and Detachment in Human–Animal Relations', *American Ethnologist* 37(2): 241–258.

Carneiro, R. 1983. 'The Cultivation of Manioc among the Kuikuru of the Upper Xingu', in R. James and W. Vickers (eds), *Adaptive Responses of Native Amazonians*. New York: Academic Press, pp. 65–111.

Clastres, P. 2010. *Archaeology of Violence*. Los Angeles: Semiotext(e).

Conklin, B., and L. Graham. 1995. 'The Shifting Middle Ground: Amazonian Indians and Eco-Politics', *American Anthropologist* 97(4): 695–710.

Descola, P. 1996. 'Constructing Natures: Symbolic Ecology and Social Practice', in P. Descola and G. Pálsson (eds), *Nature and Society: Anthropological Perspectives*. London: Routledge, pp. 82–102.

Dufour, D. 1990. 'Use of Tropical Rainforest by Native Amazonians', *Bioscience* 40(9): 652–659.

Emel, J., and J. Wolch. 1998. 'Witnessing the Animal Moment', in J. Emel and J. Wolch (eds), *Animal Geographies: Place, Politics, and Identity in the Nature-Culture Borderlands*, New York: Verso, pp. 1–24.

Fausto, C. 2007. 'Feasting on People: Eating Animals and Humans in Amazonia', *Current Anthropology* 48(4): 497–530.

———. 2008. 'Too Many Owners: Mastery and Ownership in Amazonia', *Mana* 14(2): 329–366.

Hill, C., F. Osborn and A. Plumptre. 2002. *Human-Wildlife Conflict: Identifying the Problem and Possible Solutions*. New York: Wildlife Conservation Society.

Homewood, K. Forthcoming. '"They Call it Shangri-La": Sustainable Conservation, or African Enclosures?' in M. Brightman and J. Lewis (eds), *Beyond Development and Progress: Anthropological Visions of Sustainable Futures*. New York: Palgrave.

HWCC. 2012. *Benefiting Conservation through Conflict Resolution*. Retrieved 2 May 2012 from www.humanwildlifeconflict.org.

Ingold, T. 2000. *The Perception of the Environment: Essays in Livelihood, Dwelling and Skill*. London: Routledge.

Jones, M., and A. Fabian (eds). 2006. *Conflict: The Darwin College Lectures*. Cambridge: Cambridge University Press.

Knight, J. 2003. *Waiting for Wolves in Japan: An Anthropological Study of People-Wildlife Relations*. Oxford: Oxford University Press.

Kohn, E. 2013. *How Forests Think: Towards an Anthropology Beyond the Human*. Berkeley: University of California Press.

Lagrou, E. 2000. 'Homesickness and the Cashinahua Self: A Reflection on the Embodied Condition of Relatedness', in J. Overing and A. Passes (eds), *The Anthropology of Love and Anger*. London: Routledge, pp: 152–168.

Latour, B. 1991. *Nous n'avons jamais été modernes: essai d'anthropologie symétrique*. Paris: La Découverte.

Levi-Strauss, C. 1991. *Histoire de Lynx*. Paris: Plon
McCallum, C. 2001. *Gender and Sociality in Amazonia: How Real People Are Made*. Oxford: Berg.
Nelson, A., and K. Chomitz. 2011. 'Effectiveness of Strict vs. Multiple Use Protected Areas in Reducing Tropical Forest Fires: A Global Analysis Using Matching Methods', *PLoS ONE* 6(8): e22722.
Posey, D. 1984. 'A Preliminary Report on Diversified Management of Tropical Forest by the Kayapó Indians of the Brazilian Amazon', in G. Prance and J. Kallunki (eds), 'Ethnobotany in the Tropics'. *Advances in Economic Botany* 1: 112–126.
Redford, K. 1991. 'The Ecologically Noble Savage', *Orion* 9: 24–29.
Rivière, P. 2001. 'A Predação, a Reciprocidade e o Caso das Guianas', *Mana: Estudos de Antropologia Social* 7(1): 31–53.
Shepard, G., et al. 2012. 'Hunting in Ancient and Modern Amazonia: Rethinking Sustainability', *American Anthropologist* 114(4): 652–667.
Vilaça, A. 2002. 'Making Kin Out of Others in Amazonia', *Journal of the Royal Anthropological Institute* 8(2): 347–365.
———. 2005. 'Chronically Unstable Bodies: Reflections on Amazonian Corporalities', *Journal of the Royal Anthropological Institute* 11: 445–464.
Viveiros de Castro, E. 1998. 'Cosmological Deixis and Amerindian Perspectivism', *Journal of the Royal Anthropological Institute* 4: 469–485.
Wilson, E.O. 2016. *Half-Earth: Our Planet's Fight for Life*. New York: Liveright.
Zent, S. 2012. 'Jodï Horticultural Belief, Knowledge and Practice: Incipient or Integral Cultivation?' *Boletim do Museu Parense Emílio Goeldi. Ciências Humanas* 7(2): 293–338.

6
Wildlife Value Orientations as an Approach to Understanding the Social Context of Human-Wildlife Conflict

Alia M. Dietsch, Michael J. Manfredo and Tara L. Teel

Human-wildlife conflict (HWC) is a persistent and ubiquitous problem that defies easy solution. HWC is not limited to any particular species, time or location, and it affects different people in different ways. The highly contextualized nature of HWC is largely driven by the ways in which people think about and relate to wildlife. An understanding of the impacts that human thought and action have on HWC, therefore, is critical to identifying the types of conflicts that may occur and likely human responses to those conflicts. Such an understanding can also inform conflict-mitigation efforts by identifying locations of public support for those efforts or areas where additional emphasis (e.g. communication strategies, public outreach) is needed to minimize controversy over specific management actions.

In this chapter, we describe four main assertions that stem from a programme of research conducted in the United States to inform wildlife managers of the social context influencing HWC. First, human thought and action directed at wildlife are rooted in fundamental wildlife value orientations. These wildlife value orientations are guiding ideological positions that reflect desired goals representing an 'ideal world' of human interaction with wildlife and principles for wildlife treatment. Second, wildlife value orientations are broad, enduring influences on a person that transcend time and context. Because of this, wildlife value orientations can be used to anticipate public responses (i.e. attitudes and behaviours) towards management efforts that address HWC and to identify locations where *social* conflict among different groups of people regarding the treatment of wildlife is most likely. Third, wildlife value orientations in North America are shifting at the societal level from domination to mutualism as a result of modernization, and this shift has a dramatic impact on the way in which people will respond to conflict situations and approaches for conflict remediation. Our fourth and final assertion is that HWC is most often managed at local levels, and information about the geographic distribution of wildlife value orientations can provide useful information for anticipating and managing place-based conflicts. We support each of these four assertions in a separate section of this chapter and conclude with recommendations for future research on wildlife value orientations aimed at increasing the capacity for understanding and responding to HWC in different contexts.

Wildlife Value Orientations Shape Attitudes and Behaviours towards Wildlife

Our wildlife value orientation approach builds upon a cognitive hierarchy framework commonly used in the social sciences that suggests behaviours are a function of attitudes, and attitudes are a function of more fundamental, enduring cognitions: values and value orientations (Homer and Kahle 1988). We outline these terms (i.e. attitudes, values, value orientations) here, and have provided a more thorough description of the linkages among these cognitions elsewhere (Manfredo, Teel and Henry 2009).

The Cognitive Hierarchy Framework

Pamela Homer and Lynn Kahle (1988) describe *attitudes* as positive or negative evaluative judgements of attitudinal objects (e.g. things, people, places, activities, events). Attitudes are the faster-forming cognitive processes of individuals that are ongoing and highly adaptive (Ajzen and Fishbein 1980; Homer and Kahle 1988). Attitude measures are most frequently applied in the prediction of people's behaviours, and are more accurate at doing so when the action, target, time and context are made explicit (Fishbein and Ajzen 1975; Ajzen 1988). For example, individuals with a positive evaluation of 'feeding wildlife' are more likely than those with a negative evaluation of the activity to engage in that behaviour. Furthermore, individuals with a positive evaluation of the activity 'placing a salt lick for deer in my backyard during the fall' would be more likely to do that exact behaviour than someone with a positive evaluation of the general activity of 'feeding wildlife'. Attitudes can help to explain human behaviours in relation to HWC, including individuals' engagement in activities that lead to HWC near the home (e.g. feeding deer can lead to deer damaging landscaping), and support of or opposition to different conflict-mitigation strategies.

Values are considered to be more fundamental beliefs about life goals. Values serve as principles shaping attitudes and directing behaviours through expressions of people's basic needs (Rokeach 1973; Schwartz 2006). Typically limited in number, values are culturally learned, serve as motivational constructs for individuals and transcend specific actions and situations (Schwartz and Sagie 2000). Examples of these values include honesty, respect, patriotism and personal well-being. At the cultural level, the orientation or direction of these values is shaped by ideology. Ideology is representative of the shared beliefs held by groups of people (e.g. cultures) that enable others within those groups to understand meaning, to identify who they are and to relate to one another. Ideologies are reflected in social stereotypes, principles of resource allocation, role prescriptions, origin myths, citizenship rules and other stories or ideas that define groups (Pratto 1999).

Wildlife value orientations are reflective of how cultural ideologies provide contextual meaning to the values people hold in relation to the domain of wildlife-related thought (Manfredo, Teel and Henry 2009). For example, two people may both hold the value 'be humane toward all living things', but differ in how this relates to the treatment of wildlife; one may believe that people should not harm wildlife for any reason, while the other may feel it is acceptable to kill wildlife for human purposes as long as the animal does not experience unusual pain or suffering. In this example, we contend that the values of these two individuals are guided by contrasting wildlife value orientations.

Two Primary Wildlife Value Orientations

Two primary orientations appear to be influencing human thought towards wildlife in the United States: mutualism and domination (Manfredo, Teel and Henry 2009). A domination wildlife value orientation reflects the extent to which an individual (or group of individuals) holds an ideological view of domination, or human mastery, over wildlife. Shalom Schwartz (2006) identifies mastery as a guiding principle directing individuals to actively achieve group or personal goals, even at the expense of others. People who hold a domination orientation towards wildlife are more likely to prioritize human well-being over wildlife in their attitudes and actions. The stronger a person's domination orientation, the more likely he or she is to justify the treatment of wildlife in utilitarian terms and to be accepting of actions resulting in the death of or harm to wildlife (e.g. lethal removal, trapping).

The second value orientation, mutualism, reflects the influence of an egalitarian ideology, which extends the idea of social inclusion to animals, placing an emphasis on animal equality and welfare (Wildavsky 1991). People holding a mutualism wildlife value orientation view wildlife as capable of relationships of trust with humans, as if part of an extended family, and as deserving of rights and care. The stronger one's mutualism orientation, the more likely he or she will be to engage in perceived welfare-enhancing behaviours for wildlife (e.g. feeding, nurturing abandoned or hurt animals) and to view wildlife in human terms, and the less likely he or she will be to support actions that may result in the death of or harm to wildlife.

People can then be classified into one of four types based on how strongly they identified with the item sets representing the two primary wildlife value orientations (Teel and Manfredo 2010). Utilitarians have a domination orientation towards wildlife, whereas Mutualists emphasize a mutualism orientation. Pluralists hold both a mutualism and a domination value orientation, and the influence of the two orientations is believed to be situationally contingent, meaning that the orientation playing a role in the individual's thinking on a given issue or situation will be dependent upon the exact conditions of that issue or situation (Tetlock 1986). Pluralists are likely to respond in a manner similar to that of Utilitarians for certain issues and more like Mutualists for other issues. Distanced individuals have neither a mutualism nor a domination orientation. As their label suggests, they tend to be less interested in or do not have well-formed opinions about wildlife and wildlife-related issues. The fundamental orientations (domination and mutualism) influencing these four groups offer meaningful guidance for predicting a variety of wildlife-related attitudes and behaviours.

An Understanding of Wildlife Value Orientations Can Inform Management of HWC

We assert that human responses to HWC are largely driven by people's underlying wildlife value orientations. These orientations transcend time and context, providing useful information for examining variation in beliefs regarding management of HWC. Wildlife value orientations, for example, can be used to predict levels of public acceptability for agency measures such as lethal control aimed at reducing predator-related threats to human safety (e.g. Bright, Manfredo and Fulton 2000; Teel et al. 2005; Teel and Manfredo 2010; Dietsch et al. 2011). In addition, wildlife value orientations often provide the foundation for *social* conflict among groups of people regarding how wildlife should be managed. An understanding of wildlife value orientations, therefore, can help gauge public opinion towards wildlife management strategies in the form of public support and locations of potential social conflict regarding the treatment of wildlife.

To explore the validity of wildlife value orientations across a wide range of management contexts, Teel and Manfredo (2010) overview an examination of 473 attitudinal measures in relation to the two primary wildlife value orientations.[1] The orientations were measured as composite indices of multiple, scaled items ranging from 1 'strongly disagree' to 7 'strongly agree' (see Manfredo, Teel and Henry 2009 for detail). For ease of discussion, we broadly grouped the full array of attitudinal measures into five categories related to the management and treatment of wildlife: direct and indirect harm to wildlife, wildlife/habitat protection versus human needs and interests, funding for wildlife, wildlife and human safety and provision of wildlife viewing and education opportunities (see table 6.1 for example items used in analysis). Using Pearson product-moment (or Pearson's *r*) and point biserial correlation analyses to denote statistical significance,[2] results revealed that 71 per cent (337 items) of the attitudinal measures were statistically related to domination and

Table 6.1. Examples of correlation coefficients[1] examining the relationship between wildlife value orientations and attitudinal responses to wildlife issues from a 2004 survey of western US residents.

Wildlife Issue Category *Item (two-letter state abbreviation)*	Wildlife value orientations[2]	
	Domination	Mutualism
Harm to wildlife		
Reduce the number of wolves to produce more elk and deer for hunting. (ID)	0.55	−0.34
The state wildlife agency should provide hunting opportunities. (OR)	0.65	−0.30
Wildlife/habitat protection versus human needs and interests		
Private property rights are more important than protecting declining or endangered fish and wildlife. (KS)	0.46	−0.46
Public lands should be managed to benefit species of concern even if it means providing fewer opportunities for outdoor recreation on those lands. (UT)	−0.47	0.38
Funding for wildlife		
It is acceptable to collect additional taxes for creating rural zoning regulations, which would be enforced by counties, to conserve fish and wildlife habitats. (WY)	−0.32	0.35
You should be responsible to help pay for fish and wildlife that can NOT be legally hunted, trapped or fished. (NM)	−0.38	0.37
Wildlife and human safety		
It is acceptable to destroy a mountain lion that attacks and injures a person.[3] (AZ)	0.34	−0.34
It is acceptable for the state wildlife agency to use hunters to dramatically reduce deer herds in affected zones to lower the potential spread of Chronic wasting disease. (CO)	0.47	−0.34
Provision of wildlife viewing and education opportunities		
The state wildlife agency should purchase more land for public wildlife viewing opportunities. (KS)	−0.33	0.42
The state wildlife agency should address the interests of wildlife viewers as much as it does the interests of those who hunt and fish. (AK)	−0.35	0.47

1. Unless otherwise noted, correlations presented are Pearson product-moment coefficients determined to be statistically significant (criteria defined by Bonferroni's correction [# items / 0.05] applied at the state-level; Huck 2000). Correlation coefficients ≥ 0.30 or ≤ −0.30 indicate a moderate to large effect size (Cohen 1988). Items were measured on a scale ranging from 1 'strongly disagree' or 'highly unacceptable' to 7 'strongly agree' or 'highly acceptable'.
2. Wildlife value orientations were measured as composite indices of multiple items (see Teel and Manfredo 2010 for details).
3. Correlation coefficients presented for this item are biserial and can be interpreted in the same manner as Pearson product-moment correlations (larger values indicate stronger relationships); item was measured as 0 'not acceptable' and 1 'acceptable'.

59 per cent (279 items) were related to mutualism. Of the relationships that were statistically significant, 32 per cent of correlations with domination and 31 per cent of correlations with mutualism had coefficients greater than or equal to 0.30, indicating a relationship of moderate to large practical significance (Cohen 1988). Overall, attitudinal items that were positively correlated with domination were negatively correlated with mutualism, and vice versa. In relation to the categories of wildlife-related issues, domination was most often related to items representing direct harm to wildlife (89 per cent) and least related to items representing tradeoffs between wildlife and human safety (53 per cent). For mutualism, the percentage of items within each of the categories of wildlife-related issues was highest when items represented provision of wildlife viewing and education opportunities (86 per cent) and lowest for items indicating tradeoffs between wildlife and human safety (27 per cent).

Results from these investigations support the proposal that wildlife value orientations can be used to predict responses across a variety of wildlife-related issues, and that wildlife value orientations form the basis for stark differences in human attitudes towards wildlife. As further illustration of these contrasting beliefs, Teel et al. (2010) found that mutualism in eight European countries was significantly correlated with attitudes towards wildlife conservation measures; those scoring high on mutualism were more likely to have positive attitudes towards policies that benefit wildlife. For example, German residents with a mutualism orientation were more likely than those with a domination orientation to support a ban on recreational fishing. In terms of the relationship between wildlife value orientations and wildlife-related behaviours, research has also revealed people with a domination orientation are more likely to participate in hunting and fishing, while people with a mutualism orientation are more likely to participate in wildlife observation, including bird watching (e.g. Fulton, Manfredo and Lipscomb 1996; Bright, Manfredo and Fulton 2000; Teel et al. 2005; Teel and Manfredo 2010). These examples illustrate the utility of the wildlife value orientation approach for predicting a variety of human responses across time and context.

Our investigations suggest wildlife agencies wishing to gauge the acceptability of certain conflict mitigation efforts can benefit from an understanding of wildlife value orientations. For example, we found people with a domination orientation (i.e. Utilitarians and Pluralists) to be more likely than the other wildlife value orientation types (i.e. Mutualists and Distanced individuals) to be accepting of increased hunting opportunities to control black bears in 'nuisance' HWC situations (table 6.2; Teel et al. 2010). In North America, state wildlife agencies often rely on recreational hunting as part of their management efforts for controlling species, and may use hunting as a mechanism for mitigating HWC by decreasing populations in an area where a particular species is causing localized problems (e.g. bears that create nuisance situations). State wildlife agencies, however, are increasingly finding that hunting is a controversial management practice to certain segments of their publics. Some people may see value in the use of hunting as a management tool for population control, while others may object to hunting out of a concern for the welfare of wildlife, creating a situation in which *social* conflict among groups of people over wildlife management practices may occur.

Table 6.2. Comparison of wildlife value orientation types on acceptability of providing more recreational hunting opportunities in response to black bears that are getting into trash and pet-food containers from a 2004 survey of residents living in the western United States.[1]

	Wildlife Value Orientation Type (%)					
State	Utilitarian	Pluralist	Mutualist	Distanced	X^2	Cramer's V
AK	53.7	81.2	25.9	30.6	106.34	0.45
AZ	64.6	48.8	10.5	34.0	105.29	0.47
CA	59.7	50.6	11.8	19.0	109.85	0.46
CO	71.2	52.6	16.6	28.8	138.02	0.47
HI	60.8	37.3	15.1	23.0	93.56	0.39
ID	68.1	81.0	30.3	33.9	145.07	0.43
KS	69.3	53.9	12.9	38.5	114.68	0.47
MT	65.8	82.9	24.4	52.6	179.92	0.46
ND	84.0	68.0	30.4	42.6	118.66	0.42
NE	71.9	62.8	15.4	31.5	143.62	0.47
NM	71.2	55.0	17.3	27.0	180.68	0.47
NV	65.6	52.5	13.8	20.8	129.30	0.46
OK	68.7	62.0	9.7	26.3	166.04	0.48
OR	80.2	59.7	21.0	38.8	149.04	0.50
SD	74.3	67.1	20.4	32.6	119.66	0.41
TX	60.2	44.9	16.3	29.3	71.50	0.39
UT	67.6	59.7	12.9	30.9	113.36	0.44
WA	67.6	65.6	10.6	37.5	149.50	0.53
WY	81.1	73.7	23.7	28.6	185.44	0.48

1. All X^2 values presented in this table were statistically significant ($p < 0.05$); Cramer's V values are presented as indicators of effect size, or strength of association between variables (larger values indicate stronger relationships).

Identifying social conflict 'hot spots' (i.e. locations where controversy among segments of the public regarding specific management actions is likely) can aid in the development of solutions for addressing those conflicts among groups of people. Considering the differences in attitudes of the wildlife value orientation types, our findings would suggest that the potential for social conflict over management of wildlife would be highest in places with an equal distribution of Mutualists and Utilitarians. We found this equal distribution most apparent in the states of California and Washington, which each had about a third of residents identified as Utilitarians and a third classified as Mutualists (Teel and Manfredo 2010). Wildlife agencies in these states are likely to find traditional wildlife management approaches a source for social conflict among their residents, and may benefit (e.g. gain increased public support) by focusing on the development of novel approaches for solving HWC that are more consistent with public opinion (White and Ward 2010). Agencies in locations with an equal distribution of wildlife value orientations may also gain public support for traditional mitigation efforts by targeting communication messages at people's fundamental wildlife beliefs (rather than ignoring those beliefs) in areas where support may be lacking. Our results indicate that wildlife value orientations provide a solid foundation for predicting individual level variation in attitudes and behaviours in a variety of wildlife-related contexts.

Modernization Is Causing a Shift in Wildlife Value Orientations from Domination to Mutualism

In the previous section, we laid a foundation for understanding the influence of wildlife value orientations on human attitudes and behaviours at the micro, or individual, level. We now turn our attention to the impact that macro-level forces can have on the distribution of wildlife value orientations within a society, which, in turn, can have important consequences for public responses to HWC. Manfredo, Teel and Henry (2009) have advanced the idea that modernization (i.e. greater economic well-being, rising levels of education and increased urbanization) is leading to an intergenerational shift in the way people think about wildlife in the United States. This argument regarding a shift in wildlife-related thought draws upon the work of other researchers (e.g. Inglehart 1990, 1997; Inglehart and Welzel 2005) who suggested that modernization has predictable effects on societal values. Specifically, Ronald Inglehart (1997), through longitudinal research conducted in societies around the world, has documented a shift from materialist to postmaterialist values in postindustrialized nations exhibiting greater socioeconomic status. Drawing upon psychologist Abraham Maslow's (1943) hierarchy of needs, Inglehart's theory suggests that materialist values stress the importance of fulfilling basic needs (e.g. food, clothing, economic security), while postmaterialists values focus on the fulfillment of higher-order psychological needs (e.g. belongingness, quality of life). These postmaterialist values are expected to influence other cognitive domains, including wildlife-related thought.

Building upon these ideas, Manfredo, Teel and Henry (2009) presented results obtained from hierarchical analyses of data collected in nineteen western states revealing strong, significant state-level effects. Specifically, they found that sociodemographics had a negligible impact at the micro, or individual level, but accounted for a large amount (43–77 per cent) of the variance in mean value-orientation scoring across states at the macro, or aggregate, level. Although individual variation within states obviously occurs, these results indicating variation between states cannot be ignored; modernization variables of income, education and urbanization at the state level have a significant impact on wildlife value orientations above and beyond any effect due to an individual's own level of wealth, education or size of community. To be clear, based on these findings, we would not expect an adult individual who moves to a more urbanized place or receives a sudden, significant increase in income to change his or her value orientation; however, forces of modernization impact intergenerational change at the societal level, and the value orientations of children raised under this new set of lifestyle circumstances would be impacted.

To illustrate these findings, we depict three scatter plots showing the percentage of Mutualists from each of the nineteen western states in relation to income, education and urbanization at the state level (figures 6.1, 6.2 and 6.3, respectively). The figures demonstrate that states with higher levels of income, education and urbanization have higher percentages of Mutualists. Our results are in the expected direction given Ronald Inglehart's arguments that modernization encourages post-materialist values related to belongingness needs, which is reflected in a mutualism orientation that considers animals as capable of relationships with humans and part of an extended family. It is important to note, however, that results of this specific investigation are based on cross-sectional data (i.e. data collected at a single point in time). While patterns are consistent with what we would expect if a shift due to modernization is occurring, long-term monitoring of wildlife value orientations over time is necessary to assess the validity of such a shift into the future.

A societal-level shift towards mutualism has broad implications for wildlife management. Considering the influence of value orientations on wildlife-related attitudes and behaviours described in previous sections, we would expect this shift to result in growing demand for different types of wildlife-related programmes

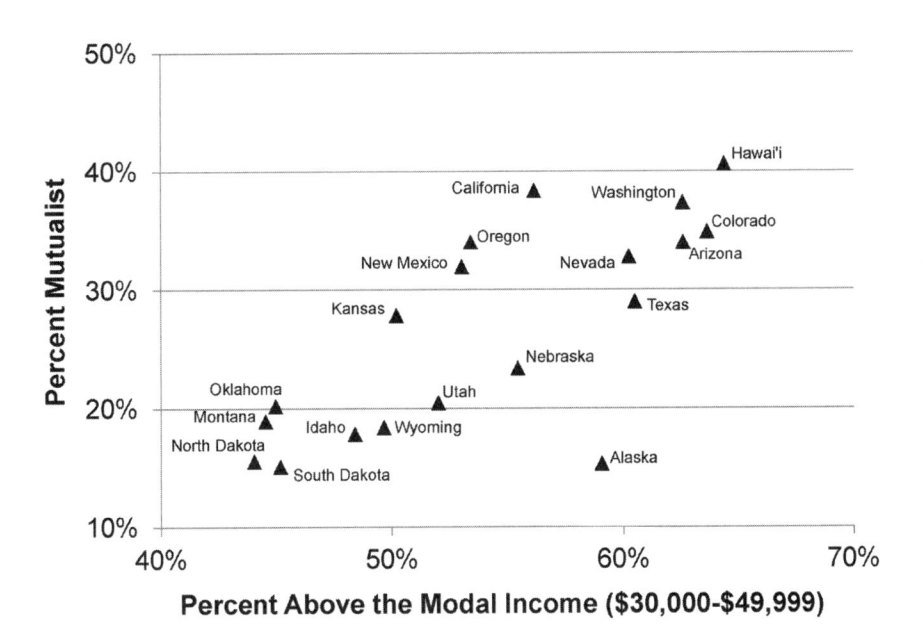

Figure 6.1. Percentage of Mutualists by income across states from a 2004 survey of residents living in the western United States. *Source*: Manfredo 2008: 207; reproduced with the kind permission of Springer Science+Business Media.

116   Understanding Conflicts about Wildlife

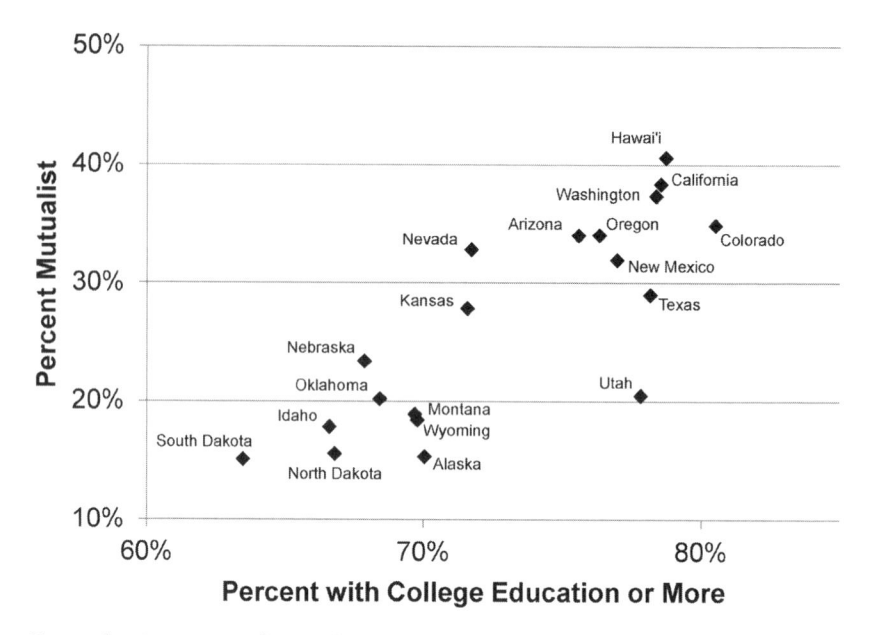

Figure 6.2. Percentage of Mutualists by education across states from a 2004 survey of residents living in the western United States.

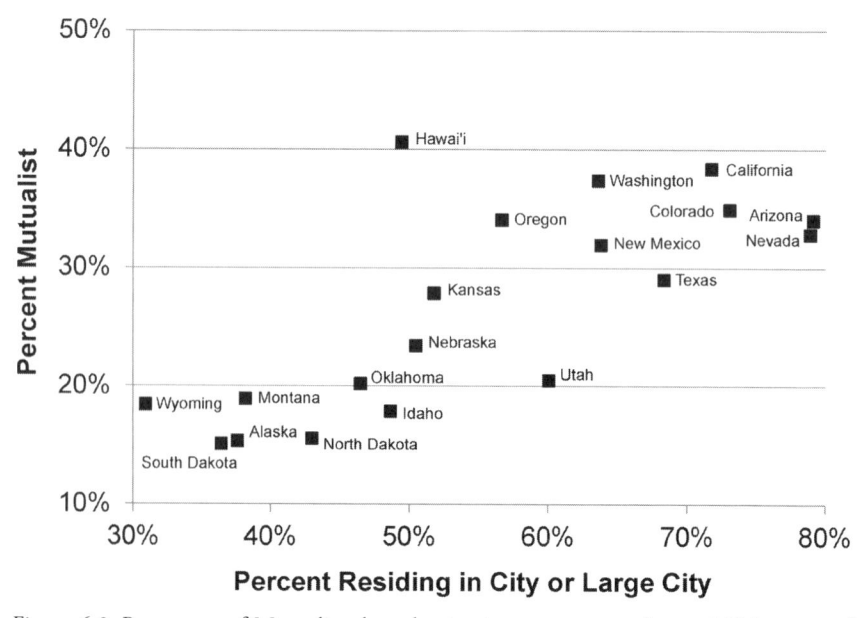

Figure 6.3. Percentage of Mutualists by urbanization across states from a 2004 survey of residents living in the western United States.

and recreational opportunities (e.g. wildlife viewing, birdwatching), and increased opposition to traditional forms of management (e.g. lethal control) and recreation (e.g. hunting, trapping). A longitudinal study conducted by the US Fish and Wildlife Service (2007) confirms that changes in wildlife-related recreation have been occurring in the United States over the last several decades.[3] State wildlife agencies dependent on the sale of hunting licenses as a significant portion of their revenue base are increasingly concerned about the documented decline in recreational hunting across much of the United States. In addition to reduced funding for wildlife management, such a decline of license sales can be a source of trouble for wildlife managers reliant on hunting for controlling certain populations of game species. For example, a decrease in hunter numbers, along with a history of predator removal, has led to localized surges in ungulate populations in certain parts of the United States and a concomitant rise in conflicts such as automobile collisions and damage to agriculture and crops (Conover 2001; Walter et al. 2010). Additionally, there have been an increasing number of lawsuits, ballot initiatives and other types of public protest against governing authorities who manage wildlife (Jacobson and Decker 2008). These types of changes in wildlife-related attitudes and behaviours create significant challenges for today's wildlife professionals who are faced with resolving HWC and managing wildlife populations with increasingly limited resources, oftentimes under circumstances that are considered controversial by various stakeholders. An understanding of the broader context of these changes, including how these changes are connected to societal shifts in public thinking about wildlife, can facilitate more proactive thinking about future management challenges and solutions.

Understanding Human Thought at Local Levels is Critical for Management of Place-Based HWC: A Case Study of Wolves in Washington

We now turn our focus to the importance of understanding human thought and action where HWC most often occurs: the local level. Consideration of the wildlife value orientation approach is increasingly useful if examined in the context of a specific community or in the location of a particular resource problem. Support for HWC alleviation efforts is largely driven by local community members who are most often affected by management decisions in addition to the types and history of conflicts in those places (Madden 2004). Knowledge of wildlife value orientations at more local levels can be useful in anticipating public responses to HWC, and can inform place-based management and outreach strategies. In recognition of this need for greater geographic specificity, we launched a multistate investigation in the United States aimed at exploring the distribution of wildlife value orientations and wildlife-related attitudes and behaviours across the landscape at different degrees of resolution (e.g. region, county, census block group). This investigation was also expected to demonstrate the utility of a spatially explicit approach for wildlife management. Here we report on an example of our findings from the state of Washington[4] to highlight the importance of using a wildlife value orientation approach in the context of place-based HWC.

118 Understanding Conflicts about Wildlife

Washington has nine wildlife commissioners appointed by the state governor, each of whom represents residents living in a particular area of the state. At least three of these commissioners must reside east of the Cascade Mountains, while at least three others must reside west of this physical divide. No two commissioners can reside in the same jurisdiction. These wildlife commissioners are increasingly confronted with addressing and responding to social conflict among the state's residents regarding the appropriateness of different wildlife management approaches. The difficulty facing these commissioners in resolving conflicts becomes apparent when the distribution of wildlife value orientations is depicted by county for the entire state (figure 6.4). Counties in the eastern portion of the state have a greater number of Utilitarians than Mutualists (in some cases, over 40 per cent more), while counties in the northwestern portion of the state have a greater number of Mutualists than Utilitarians. Washington's wildlife commissioners are, in essence, tasked with

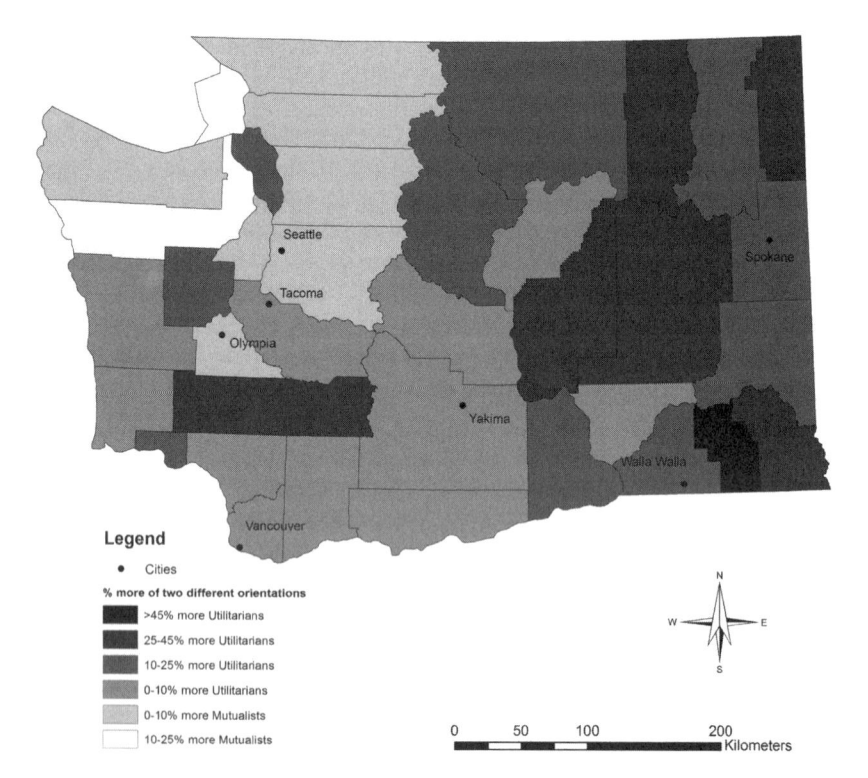

Figure 6.4. Distribution of two wildlife value orientation types (i.e. Mutualists or Utilitarians) by county from a 2009 survey of Washington residents. Map represents the percentage of residents within each of Washington's thirty-nine counties that were classified as one value orientation (Utilitarians) subtracted from the percentage of residents classified as the other value orientation (Mutualists) to determine the primary orientation found within that county and the degree to which ('per cent more') that orientation exists.[1]

representing residents who hold contrasting wildlife value orientations and have different expectations regarding how wildlife should be managed.

The dissention among Washington residents over wildlife-related issues can be demonstrated through an example involving the re-establishment of wolves into the state. Wolves have been extirpated for seventy years and are returning to Washington from surrounding states and Canadian provinces. At the time of the study, wolves were fully protected as endangered under state law (RCW77.15.120) and protected under the federal Endangered Species Act in the western two-thirds of the state (Washington Department of Fish and Wildlife 2011), meaning that unlawful take (e.g. killing) of wolves in Washington was punishable by law. While some residents are excited about wolves returning to the state and the prospect of seeing wolves in the wild, other residents are concerned about human-wolf conflicts and how to mitigate conflicts that may occur. Example conflicts include wolf damage to human property (e.g. livestock, pets) or situations in which humans feel their own safety is threatened by wolves. To ensure that wolf recovery goals are met by the state wildlife agency, potential conflicts between residents and wolves must be quickly addressed in ways that foster enduring public support while allowing wolves a chance to establish and thrive.

Results from our investigation indicate a high level of statewide support for wolf recovery, but considerable variation in wolf-related beliefs across counties. This variation was consistent with the distribution of wildlife value orientations; support for wolf recovery was greatest in areas where there were more Mutualists and lowest in areas where there were more Utilitarians. Mutualists, however, were predominately found in northwest Washington, the location of the urbanized area of Seattle and more than half of the state's residents. Statewide results indicating overwhelming support for wolf recovery, therefore, would be driven by residents living in areas that likely experience few, if any, wolf-related conflicts. In contrast, residents in the geographically vast and more rural eastern portion of the state constitute a smaller portion of the state's overall population and would be represented accordingly in a statewide sample, yet these are areas where human-wolf interactions are more likely. For example, at the time of our study, the state wildlife agency confirmed the establishment of two wolf packs in the central portion of the state and three wolf packs in the northeast, far from counties where support for wolf recovery was found (figure 6.5). This support (or lack thereof) for wolf recovery stems from people's underlying values, and can form the basis for social conflict over statewide management of wolves.

Our data indicate that public sentiment regarding mitigation of potential human-wolf conflicts is also consistent with the countywide distribution of wildlife value orientations in Washington. Residents from counties in northeastern Washington (a Utilitarian stronghold) were supportive of more invasive wolf management techniques, while residents of counties near the urban area of Seattle (a Mutualist stronghold) found the same three techniques to be less acceptable. These techniques included: limiting the number of wolves if they cause declines in deer and elk populations in certain areas, lethal removal of wolves known to have caused loss of livestock and establishing a recreational hunt of wolves once state recovery goals are

120 Understanding Conflicts about Wildlife

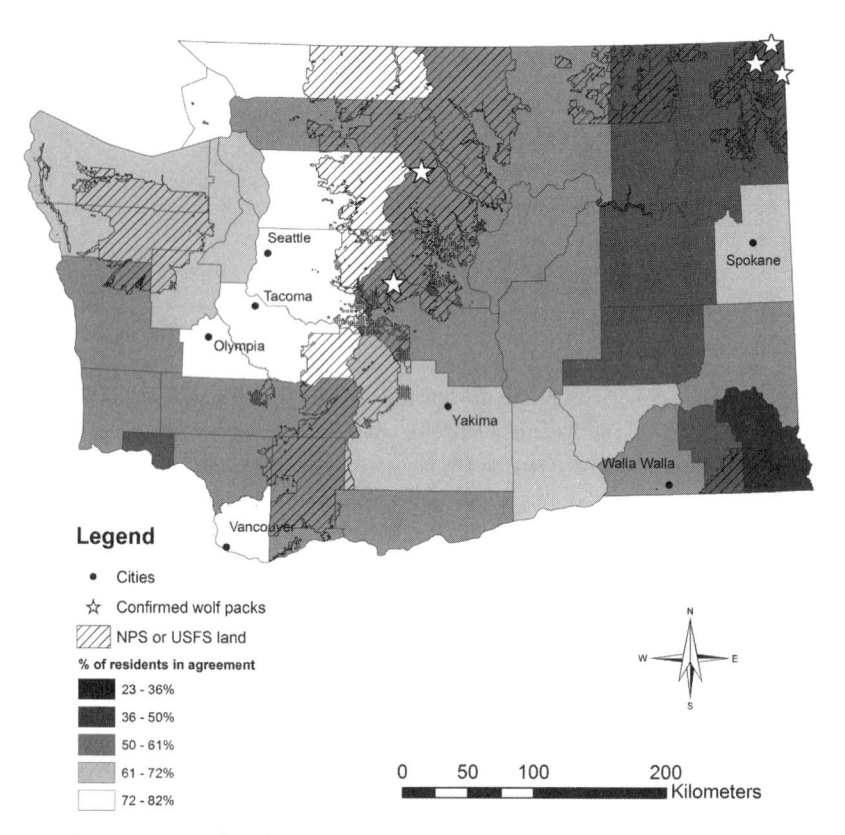

Figure 6.5. Percentage of residents by county accepting of wolves being moved from one area to another to help establish new wolf populations from a 2009 survey of Washington residents. Map has an overlay of large tracts of federal land (i.e. National Park Service [NPS] and US Forest Service [USFS] lands) representing habitat conducive to wolf establishment. Wolf packs confirmed by the state wildlife agency, Washington Department of Fish and Wildlife, to be established within the state at the time of the survey are also indicated.

reached. These results suggest that residents in the eastern region are likely to expect the agency to use lethal control as a management strategy, while residents in other regions of the state may balk at such measures, creating a difficult situation for wildlife managers balancing human needs with healthy wolf populations within the state.

Information about the distribution of wildlife value orientations at finer degrees of resolution provides a useful guide for informing local-level solutions to HWC that are oftentimes dependent upon public support (Dietsch, Teel and Manfredo 2016), as well as the nature of the problem or resource conditions in a particular location. Our investigation in Washington, for example, pinpointed locations where human tolerance of wolf recovery is low, and demonstrated that variation in tolerance levels across the state is linked to the countywide distribution of wildlife value orientations. Since wildlife value orientations have been shown to have predictive validity in gauging public response across a host of wildlife-related issues, we would expect that portions of our findings could be extended to other HWC scenarios. To illustrate, counties where residents primarily hold a mutualism orientation would be less likely to support lethal control as a means to reduce conflicts with other types of predator species (e.g. black bears, mountain lions), while counties where residents primarily hold a domination orientation would be more likely to support this mitigation strategy.

Managers and practitioners can clearly benefit from a local level understanding of wildlife value orientations. However, it is important to note that wildlife value orientations alone are not enough to predict responses to all potential human-wildlife interaction scenarios (Dietsch, Teel and Manfredo 2016). Additional sources of data, including coupling of biological data (e.g. wolf pack locations, human-wildlife conflict hot spots) with social data such as values and attitudes, will provide a more holistic tool from which managers and practitioners can address place-based conflict.

Conclusion

Human-wildlife conflict is a complex issue that can have serious impacts on human livelihoods and consequences for the survival of species in many areas (IUCN 2003; DiStefano 2005). Due to these multifarious concerns, much has been published over the last several decades on HWC, with a variety of viable mitigation strategies offered. This literature, however, often overlooks the social drivers contributing to HWC or has indicated failure to resolve conflicts long-term due to a lack of attention to these human influences. Researchers (e.g. Treves and Karanth 2003; Madden 2004; Dickman 2010) are increasingly calling for a focus on collaborative, multi-disciplinary efforts that aim to understand the impact of human thought and action on HWC, as such an understanding is critical to designing appropriate and effective conflict-alleviation strategies.

Our theoretical approach to investigating human influences on HWC draws upon the social sciences, and more specifically, on the cognitive hierarchy framework from social psychology. We assert that fundamental wildlife value orientations shape human attitudes and behaviours on a variety of wildlife-related issues. We have identified two contrasting wildlife value orientations – domination and mutualism – that are reflective of broader cultural ideologies found in the United States, and explained how these orientations form the basis of human thought about and responses to wildlife. A value orientation of domination reflects an ideology of human mastery over wildlife, while mutualism extends an egalitarian ideology to include wildlife. Our work is intended to provide a clear conceptual framework for understanding diverse viewpoints on wildlife issues, and to distinguish among various types of cognition that form the basis for human behaviour. Each level of cognition has different implications for wildlife management. For example, values and value orientations form early in life and are not likely to change much once formed, while attitudes (the more immediate antecedents to behaviour) are less stable and, therefore, more likely to be influenced by targeted communication efforts (Teel and Manfredo 2010).

We have primarily focused on our research that explores the influence of values on wildlife-related attitudes and behaviours considering the cultural context of US society. Wildlife management and conservation in North America have been heavily influenced by a strong hunting tradition (e.g. Organ and Fritzell 2000), and the existence of a domination orientation within the United States is consistent with such a tradition. We have also explained a rise in mutualism within the United States, based on documented increases in postmaterialist values stressing belongingness needs in countries exhibiting higher socioeconomic status (Inglehart 1997). Wildlife value orientations, which are shaped by underlying cultural ideologies, may be quite different in other countries and could impact the direction of human attitudes and behaviours towards wildlife in ways that are different than those we found in US society today (Manfredo and Dayer 2004). The Chinese, for example, have relied heavily on the use of wildlife parts (e.g. rhinoceroses' horns, tiger bones) in traditional medicines for thousands of years (Mainka and Mills 1995); these practices are rooted in a very strong utilitarian tradition (Harris 2008). As China becomes increasingly prosperous and exhibits increased levels of modernization, we would not expect the same increase in mutualism-related thought as found in the United States. Therefore, an exploration of the wildlife value orientation concept cross-culturally is necessary for validating the utility of our work and building theory (Teel, Manfredo and Stinchfield 2007). Understanding wildlife value orientations in different cultural contexts will likely prove to be as useful as it has in the United States for predicting human responses to wildlife-related topics, including HWC.

In response to this need for understanding wildlife value orientations cross-culturally, a series of investigations was launched. Exploratory, qualitative studies of wildlife value orientations have been employed in various countries, including the Netherlands, China, Estonia, Mongolia and Thailand as part of the Wildlife Values Globally Research Program (see *Human Dimensions of Wildlife*, volume 12, issue 5). Wildlife value orientations have also been examined cross-culturally through a quantitative study in ten European countries (Teel et al. 2010). While these research efforts have helped to detect domination and mutualism orientations within different societies, continued work is necessary to uncover the full array of wildlife value orientations and the impacts these orientations have on attitudes and behaviours in the context of HWC. In addition, there is a need for further research exploring how wildlife value orientations may vary among subcultures (e.g. those with different ethnicities or ancestries, immigrant populations) *within* societies (Manfredo, Teel and Dietsch 2016). For example, our work supports the notion that ancestry plays a role in shaping values, and that the domination wildlife value orientation found in the United States appears to have originated out of European Judeo-Christian traditions (Manfredo, Teel and Dietsch 2016). Will this influence remain in areas along the American–Mexican border where Hispanic populations are increasing (US Census Bureau 2011)? The success of wildlife management strategies there may depend on managers' abilities to understand and engage Hispanic residents as stakeholders in wildlife issues (Lopez et al. 2005).

Our investigations reveal that information about human thought and behaviour can enhance understanding of the social context of HWC. Such information may include factors that are at the root of HWC problems as well as factors that can affect the long-term success of HWC management interventions. Wildlife value orientations offer a useful and parsimonious approach for describing how different people think about wildlife and the impacts that this can have on their reactions to wildlife-related issues. State wildlife agencies that have participated in our research efforts have taken a critical step in developing baseline information regarding the diversity of public interests within their states and identifying potential sources of public controversy over their wildlife management efforts. Future research applications that can more fully examine the variation in wildlife value orientations and associated implications for HWC across different geographic, temporal, cultural and ecological scales offer promise in being able to inform conflict mitigation strategies that more readily account for the social context of HWC.

Alia M. Dietsch is an Assistant Professor in the School of Environment and Natural Resources at the Ohio State University. Her research, teaching, and outreach efforts focus on the role of social science theories and methodologies in improving conservation decision-making and natural resource, parks and protected-area management. She primarily uses a social psychological approach to understand the influence of human cognition on behavior across various multilevel and socioecological contexts. Her work is also closely aligned with state and federal natural resource agencies' efforts to enhance biodiversity conservation through strategic planning, management and communication.

Michael J. Manfredo is a Professor and Head of the Department of Human Dimensions of Natural Resources at Colorado State University. His research, teaching and outreach activities focus on applying attitude and value theory to natural resource issues and advancing social science interdisciplinarity. He has published over eighty-five peer-reviewed articles and co-founded the journal *Human Dimensions of Wildlife*. He has co-edited several books, including the award-winning *Wildlife and Society: The Science of Human Dimensions*. He has been principal investigator on over eighty research projects, including investigating wildlife values across the United States.

Tara L. Teel is an Associate Professor in the Department of Human Dimensions of Natural Resources at Colorado State University. Her work focuses on improving conservation decision-making through understanding human thought and behavior, and through building social science capacity among conservation professionals. Much of her research has been devoted to examining human-wildlife relationships in the United States and globally, with a particular focus on employing concepts and methodologies from social psychology to study human values towards wildlife, attitudes and behaviors towards wildlife-related issues, and the social factors underlying human-wildlife conflict.

Notes

1. Data presented throughout this chapter, unless indicated otherwise, were collected as part of a 2004 survey of residents in the western United States (*n* = 12,673) entitled *Wildlife Values in the West* (Teel et al. 2005).
2. Criteria for statistical significance were defined using Bonferroni's correction (# items / .05) applied at the state-level (Huck 2000).
3. A recent study by the U.S. Fish and Wildlife Service (2012) indicates that participation in recreational hunting has increased 5 per cent from 2001 to 2011; however, it is too soon to tell what factors may have affected such a change and whether declines documented over previous decades have ended. For example, increased 'leisure' time and reduced income levels due to job losses occurring in the United States as part of the 2008 economic downturn may have led to a temporary stabilization of hunting participation rather than a reversal of a decades-long decline in participation.
4. Data presented in this section were collected as part of a 2009 survey of Washington residents (*n* = 4,183) entitled *Understanding People in Places* (Dietsch et al. 2011).

References

Ajzen, I. 1988. *Attitudes, Personality, and Behaviour*. Milton-Keynes: Open University Press.

Ajzen, I., and M. Fishbein. 1980. *Understanding Attitudes and Predicting Social Behaviour*. Englewood Cliffs, NJ: Prentice-Hall.

Bright, A.D., M.J. Manfredo and D.C. Fulton. 2000. 'Segmenting the Public: An Application of Value Orientations to Wildlife Planning in Colorado', *Wildlife Society Bulletin* 28(1): 218–226.

Cohen, J. 1988. *Statistical Power Analysis for the Behavioural Sciences*, 2nd ed. Hillsdale, NJ: Lawrence Erlbaum Associates.

Conover, M.R. 2001. *Resolving Human-Wildlife Conflicts: The Science of Wildlife Damage Management*. Boca Raton, FL: CRC Press.

Dickman, A.J. 2010. 'Complexities of Conflict: The Importance of Considering Social Factors for Effectively Resolving Human–Wildlife Conflict', *Animal Conservation* 13: 458–466.

Dietsch, A.M., et al. 2011. *State Report for Washington from the Research Project Entitled 'Understanding People in Places'*. Project Report for the Washington Department of Fish and Wildlife. Fort Collins: Colorado State University, Department of Human Dimensions of Natural Resources.

Dietsch, A.M., T.L. Teel and M.J. Manfredo. 2016. 'Social Values and Biodiversity Conservation in a Dynamic World', *Conservation Biology*. doi:10.1111/cobi.12742.

DiStefano, E. 2005. *Human–Wildlife Conflict Worldwide: Collection of Case Studies, Analysis of Management Strategies and Good Practices*. Rome: SARD Initiative Report.

Fishbein, M., and I. Ajzen. 1975. *Belief, Attitude, Intention, and Behaviour: An Introduction to Theory and Research*. Reading, MA: Addison-Wesley.

Fulton, D.C., M.J. Manfredo and J. Lipscomb. 1996. 'Wildlife Value Orientations: A Conceptual and Measurement Approach', *Human Dimensions of Wildlife* 1(2): 24–47.

Harris, R.B. 2008. *Wildlife Conservation in China: Preserving the Habitat of China's Wild West*. Armonk, NY: M.E. Sharpe, Inc.

Homer, P.M., and L.R. Kahle. 1988. 'A Structural Equation Test of the Value–Attitude–Behaviour Hierarchy', *Journal of Personality and Social Psychology* 54(4): 638–646.

Huck, S.W. 2000. *Reading Statistics and Research*, 3rd ed. New York: HarperCollins.
Inglehart, R. 1990. *Culture Shift in Advanced Industrial Society*. Princeton, NJ: Princeton University Press.
———. 1997. *Modernization and Postmodernization*. Princeton, NJ: Princeton University Press.
Inglehart, R., and C. Welzel. 2005. 'Exploring the Unknown: Predicting the Responses of Publics Not Yet Surveyed', *International Review of Sociology* 15(1): 173–201.
International Union for Conservation of Nature (IUCN). 2003. *Recommendations*. Vth IUCN World Parks Congress (8–17 September 2003). Durban, South Africa. WPC Rec 5.01.
Jacobson, C.A., and D.J. Decker. 2008. 'Governance of State Wildlife Management: Reform and Revive or Resist and Retrench?' *Society and Natural Resources* 21: 441–448.
Lopez, R.R., et al. 2005. 'Changing Hispanic Demographics: Challenges in Natural Resource Management', *Wildlife Society Bulletin* 33(2): 553–564.
Madden, F. 2004. 'Creating Coexistence between Humans and Wildlife: Global Perspectives on Local Efforts to Address Human-Wildlife Conflict', *Human Dimensions of Wildlife* 9(4): 247–257.
Mainka, S.A., and J.A. Mills. 1995. 'Wildlife and Traditional Chinese Medicine: Supply and Demand for Wildlife Species', *Journal of Zoo and Wildlife Medicine* 26(2): 193–200.
Manfredo, M.J. 2008. *Who Cares about Wildlife: Social Science Concepts for Exploring Human-Wildlife Relationships and Conservation Issues*. New York: Springer-Verlag Press.
Manfredo, M.J., and A.A. Dayer. 2004. 'Concepts for Exploring the Social Aspects of Human-Wildlife Conflict in a Global Context', *Human Dimensions of Wildlife* 9(4): 1–20.
Manfredo, M.J., T.L. Teel and A.M. Dietsch. 2016. 'Implications of Human Value Shift and Persistence for Biodiversity Conservation', *Conservation Biology* 30(2): 287–296. doi:10.1111/cobi.12619.
Manfredo, M.J., T.L. Teel and K.L. Henry. 2009. 'Linking Society and Environment: A Multi-Level Model of Shifting Wildlife Value Orientations in the Western U.S.', *Social Science Quarterly* 90(2): 407–427.
Maslow, A.H. 1943. 'A Theory of Human Motivation', *Psychological Review* 50(4): 370–396.
Organ, J.F., and E.K. Fritzell. 2000. 'Trends in Consumptive Recreation and the Wildlife Profession', *Wildlife Society Bulletin* 28(4): 780–787.
Pratto, F. 1999. 'The Puzzle of Continuing Group Inequality: Piecing Together Psychological, Social, and Cultural Forces in Social Dominance Theory', in M.P. Zanna (ed.), *Advances in Experimental Social Psychology*, vol. 31. San Diego, CA: Academic Press, pp. 191–263.
Rokeach, M. 1973. *The Nature of Human Values*. New York: The Free Press.
Schwartz, S.H. 2006. 'A Theory of Cultural Value Orientations: Explication and Applications', *Comparative Sociology* 5(2–3): 137–182.
Schwartz, S.H., and G. Sagie. 2000. 'Value Consensus and Importance: A Cross-National Study', *Journal of Cross-Cultural Psychology* 31(4): 465–497.
Teel, T.L., et al. 2005. *Regional Results from the Research Project Entitled 'Wildlife Values in the West'* (Project Rep. No. 58). Project Report for the Western Association of Fish and Wildlife Agencies. Fort Collins: Colorado State University, Human Dimensions in Natural Resources Unit.
———. 2010. 'Understanding the Cognitive Basis for Human-Wildlife Relationships', *International Journal of Sociology* 40(3): 104–123.
Teel, T.L., and M.J. Manfredo. 2010. 'Understanding the Diversity of Public Interests in Wildlife Conservation', *Conservation Biology* 24(1): 128–139.

Teel, T.L., M.J. Manfredo, and H.M. Stinchfield. 2007. 'The Need and Theoretical Basis for Exploring Wildlife Value Orientations Cross-Culturally', *Human Dimensions of Wildlife* 12(5): 297–305.

Tetlock, P.E. 1986. 'A Value Pluralism Model of Ideological Reasoning', *Journal of Personality and Social Psychology* 50(4): 819–827.

Treves, A., and U. Karanth. 2003. 'Human–Carnivore Conflict and Perspectives on Carnivore Management Worldwide', *Conservation Biology* 17(6): 1491–1499.

U.S. Census Bureau. 2011. 'Overview of Race and Hispanic Origin: 2010, 2010 Census Briefs', Washington, DC. Retrieved 7 March 2012 from http://www.census.gov/prod/cen2010/briefs/c2010br-02.pdf.

US Fish and Wildlife Service. 2007. *2006 National Survey of Fishing, Hunting, and Wildlife-Associated Recreation: National Overview*. Washington, DC: US Department of the Interior.

———. 2012. *2011 National Survey of Fishing, Hunting, and Wildlife-Associated Recreation: National Overview*. Washington, DC: US Department of the Interior.

Walter, W.D., et al. 2010. 'Management of Damage by Elk (*Cervus elaphus*) in North America: A Review', *Wildlife Research* 37: 630–646.

Washington Department of Fish and Wildlife. 2011. 'Wolves in Washington' (fact sheet). Retrieved 1 March 2012 from http://wdfw.wa.gov/conservation/gray_wolf/gray_wolf_fact_sheet.html.

White, P.C.L., and A.I. Ward. 2010. 'Interdisciplinary Approaches for the Management of Existing and Emerging Human-Wildlife Conflicts', *Wildlife Research* 37: 623–629.

Wildavsky, A.B. 1991. *The Rise of Radical Egalitarianism*. Washington, DC: American University Press.

7
A Long-Term Comparison of Local Perceptions of Crop Loss to Wildlife at Kibale National Park, Uganda
Exploring Consistency Across Individuals and Sites

Lisa Naughton-Treves, Jessica L'Roe, Andrew L'Roe and Adrian Treves

Wildlife crop-raiding often occurs in a patchy distribution within forest-agricultural landscapes. Some analysts blame conflict hotspots on individual animals, e.g. an elephant develops a taste for sugar cane; others fault farmers for planting fields in prime wildlife habitat. Beyond the behaviour of 'problem' animals or people, ecologists point to a certain mix of forest and agricultural habitat for creating conflict hotspots (Nyhus, Tilson and Sumianto 2000; Chiyo et al. 2005). The timing of crop-raiding is also clumped in episodes associated with crop or forest phenology or patterns of wildlife movement (Chiyo et al. 2005). Such spatial and temporal variation makes it hard to assess the severity of crop loss in a given location and whether it is changing over time. Disentangling explanatory factors for conflict and deciphering their respective weight is important for designing effective interventions. Such detective work is important given that human-wildlife conflict provokes controversy and may undermine support for conservation. Systematic measurement of crop loss amounts and patterns is important for effective management, as is better understanding of social factors creating and amplifying perceptions of risk among local residents. The patchy nature of crop-raiding (in space and time) presents methodological challenges for measuring and predicting conflict, particularly in the highly heterogeneous agroecosystems of tropical small-scale agriculture.

A second source of complexity concerns people's perceptions of wildlife and crop loss. Several studies conclude that perceived risk of loss is as important as actual losses, and that these perceptions are shaped by both social and ecological factors (Ogra 2008; Dickman 2010). When a local farmer is asked 'what's the worst crop-raiding species?' her answer is filtered through recall of not only her own experience but that of her neighbours, plus her feelings about the park, wildlife authorities and even the researcher posing the question. Larger, more public discussions of the problem yield different descriptions than private conversations (Naughton-Treves 1997). Changes in wildlife numbers and behaviours complicate assessments of risk and tolerance, as do policy changes. For example, conflict complaints may surge when compensation programs are launched (Gachago and Waithaka 1995). Similarly, people may resent raiding wildlife species if new conservation rules constrain their traditional methods for defending crops.

In this chapter we explore how perceptions of 'problem' wildlife change over time among citizens neighbouring Kibale Forest in Uganda. Our study period spans seventeen years after Kibale was upgraded from a reserve to a park, a change that provoked concern among local citizens about increased crop losses to wildlife. By canvassing all farmers in the same park-edge communities and using the same survey techniques, we hope to detect change in perceptions across time. A longitudinal panel also allows us to explore constancy in complaints about wildlife, whether by comparing responses by the same individual in two time periods, or responses of different individuals residing on the same site at different times. Thus we hope to add to the broader literature regarding the stability of environmental attitudes and factors associated with change. Environmental sociologists point out that attitudes are shaped by people's underlying values and these values are likely to change slowly, if at all (Manfredo 2008; Heberlein 2012). Identifying causal explanations for attitude change is challenging and depends on the scale of analysis. Some scholars emphasize that direct personal experience can change individual attitudes; others point to shifts in public discourse or long-term socioeconomic change to explain broader attitude change (Teel, Manfredo and Stinchfield 2007). Finding causal explanations is all the more difficult without longitudinal data on individual attitudes. In the realm of wildlife, many important studies of attitude change have employed cross-sectional approaches (Teel and Manfredo 2010). The reintroduction and/or recovery of large carnivores has attracted special attention and yielded somewhat contradictory results regarding stability of attitudes and, when attitudes do change, explanatory factors. For example, residents of Scandinavia and Minnesota in the United States became more positive towards wolves over time as their fear decreased and/or familiarity increased (Kellert 1999; Zimmerman, Wabakken and Dotterer 2001). But subsequent, larger samples from Sweden found attitudes towards wolves became more negative after wolves returned (Ericsson and Heberlein 2003). Residents of Utah expressed similar attitudes towards wolves before and after the species was restored in nearby states (Bruskotter, Schmidt and Teel 2007). Aleksandra Majić et al. (2011) working in Croatia found reduced acceptance of brown bears (*Ursus arctos*) despite no change in reported livestock depredations. For each study above, the causes of reported changes were difficult to interpret without knowledge of how individuals changed over time. For example, Melanie Houston, Jeremy Bruskotter and David Fan (2010) showed that attitudes about wolves changed over time, but they could not distinguish between changes in attitudes and changes in cohorts of reporters, respondents, etc. All studies that attempt to detect attitude change via cross-sectional approaches are challenged to disentangle demographic change over time from individual attitude change. By contrast, longitudinal data on individuals (panels of respondents resampled at two or more time points) can distinguish individual histories that may help explain change or detect if wildlife management interventions affected attitudes (Henderson et al. 2000). In short, longitudinal studies provide stronger inferences about the causes of change in individual attitudes and who has changed. A panel study of several hundred residents within the wolf range in Wisconsin showed significant increase in negative feelings towards wolves over time (Treves, Naughton-Treves and Shelley 2013). Interestingly, individual experience showed no relation to attitude change – i.e. those suffering wolf attacks on their livestock during the study period did not become more resentful or fearful of wolves than their counterparts free of such conflict.

Few long-term panel studies on attitudes towards wildlife have been published, especially from Africa. Here we explore the constancy of attitudes towards crop-raiding species among small-scale agriculturalists neighbouring a Ugandan forest. Although we were not able to measure certain key variables (e.g. wildlife population dynamics over seventeen years), we explore constancy in wildlife complaint comparing responses by the same individual in two time periods versus responses of different individuals. The fact that we controlled for respondents' locations at a fairly fine resolution (~1 ha) is noteworthy. Controlling for location is important at our study site because the risk of crop loss to wildlife drops off steeply over a few 100 m from the park edge, and there seem to be localized hotspots of conflict. Finally, of potential interest to wildlife managers, this longitudinal approach also allows us to test whether standard predictors of perceived raiding intensity persist, e.g. a respondent's proximity to reserve, ethnicity, sex and village location (Naughton-Treves 1998).

Study Site and Background

Kibale National Park is a 760 km^2 forest remnant located in western Uganda (figure 7.1). Kibale is rich in primates and other species (Struhsaker 1997), including those notorious for feeding on crops, such as olive baboons (*Papio cynocephalus*), red-tail monkeys (*Cercopithecus ascanius*), elephants (*Loxodonta africana*) and bushpig (*Potamochoerus* sp.). At the start of our study (1994), ~54 per cent of the land within 1 km of Kibale's western boundary was used for smallholder agriculture (Mugisha 1994). Historically and today, farmers in the area belong to two predominant ethnic groups, the long-present Toro, and the Kiga, who came to the region during the

1950s and 1960s (Turyahikayo-Rugyema 1974). Toro chiefs traditionally allocated land to immigrants on the outskirts of their settlements, in part to buffer Toro farmers from crop damage by wildlife (Aluma et al. 1989). Both groups plant more than thirty species of subsistence and cash crops including maize, beans, yams, sweet potatoes, cassava and *Musa* spp. (bananas and plantains). Farm sizes are small, aver-

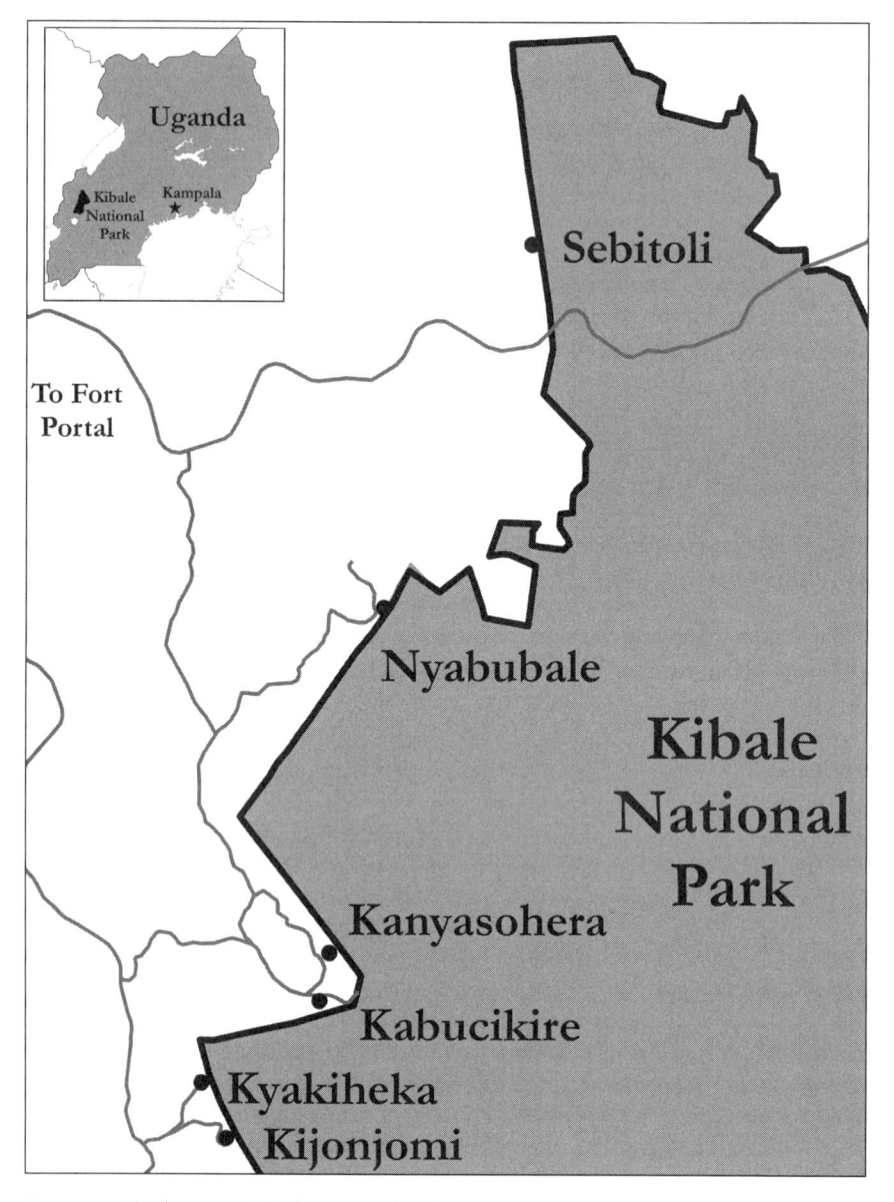

Figure 7.1. Study sites at Kibale National Park, Uganda.

aging 1.4 ha at the start of the study in 1994, and population density is high at 272 individuals per km².[1] Within this diverse farming system, various wildlife species make forays from the park to feed on crops (for more on wildlife conflicts at Kibale, see Naughton-Treves 1997; MacKenzie and Ahobyona 2012).

In 1993, the Kibale Forest Reserve was regazetted as a 760 km² national park (NEMA 1997). Systematic data on park-level wildlife populations are scarce, but surveys have indicated that total elephant numbers have been increasing since the period of illegal poaching and control shooting in the 1970s and 1980s (Struhsaker 1997), to an estimated 262 animals in 2000 (Plumptre et al. 2001) and to nearly 400 in 2005 (Wanyama et al. 2010). Farmers living on Kibale's edge have long complained about crop loss to animals and tend to express their frustration most intensely during group discussions (Naughton-Treves 1997). During the early years of study people asked, 'Why should we starve so that baboons may eat?' In response to such complaints around Kibale and other parks, in 2000, the Uganda Wildlife Authority authorized the 'supervised' hunting of three species designated vermin: baboons, bushpigs and vervet monkeys (*Cercopithecus aethiops*) (Olupot, McNeilage and Plumptre 2009).

Insights from Previous Physical Measurement of Crop-Raiding at Kibale

During 1992–1994, we monitored crop damage to animals weekly in six villages lying within 500 m of Kibale's boundary (Naughton-Treves 1998) and again in 1997 (for six months in three villages only on a biweekly basis) (figure 7.1). We canvassed all farms and recorded damage by twelve varieties of animal (baboon, bird, bushbuck, bushpig, cattle, chicken, chimpanzee, civet, domestic pig, elephant, giant rat, goat, mouse, redtail monkey). Three key results from the systematic measurement during these years are: the amount and frequency of crop loss varied markedly within villages, between villages and between species; the strongest predictor of damage was proximity to the park boundary; 90 per cent of damage events occurred within 160 m of the forest boundary.

Five wildlife species accounted for 85 per cent of the forays into fields: baboons, bushpigs, redtail monkeys, chimpanzees and elephants. Pooling the data for six villages, redtail monkeys were the most frequent crop foragers and baboons caused the most cumulative damage. Elephant raiding was concentrated in certain villages, and on certain farms within these villages. Farmers' individual defensive strategies (e.g. guarding, snaring) diminished damage by some species (namely bushpigs), but did not appear to deter elephants.

Underlying Predictions for Longitudinal Study

Our investigation is organized around two basic themes. First, we use crop-raiding complaints and 'worst animal' rankings in the two sample periods to draw inference about changing patterns of crop loss at Kibale's edge. Second, we use our data to explore relative constancy in attitudes by comparing panel data to cross-sectional data. Specific predictions for these two themes are:

Theme One: Inference about Crop Loss to Wildlife over Time

Prediction 1: Crop-raiding complaints will increase and extend further from the park edge over time.

Or, by contrast:

Prediction 2: Crop-raiding complaints over time will be increasingly confined to the park edge due to the depletion of wildlife habitat and increased human density around the park.

These two predictions are partly contradictory because there are numerous site-specific factors other than human population density or protected-area status that shape local patterns and perceptions of crop-raiding. Indeed, we test three other factors: the respondent's village, ethnicity and sex, because all three proved significant in previous analyses (Naughton-Treves 1997). We justify our Theme One predictions and emphasis on distance from park as follows: Kibale's upgrade to park status in 1993 was intended to improve wildlife protection. Thus we expect more abundant wildlife and more widespread raiding incidents. On the other hand, human population density around Kibale has increased since 1994. Other studies show an inverse relationship between human population density and the distance wildlife travel to raid crops beyond the park boundary (Newmark 1996). Finally, distance to park has consistently proved a significant predictor of crop loss in other studies at Kibale (Hartter, Goldman and Southworth 2011; Mackenzie 2012).

Theme Two: Exploring the Consistency of Attitudes over Time

Prediction 1: Reported crop-raiding intensity and problem-animal rankings will vary less when the same respondent is interviewed in T1 (1994) and T2 (2011–2012).

Prediction 2: Reported crop-raiding intensity and problem-animal rankings will vary less when T1 and T2 respondents are family members versus strangers with one another.

For both these predictions, we assume that we can control for other factors shaping perceived patterns of crop loss by matching T1 and T2 interviews by location, i.e. we tried to conduct repeat interviews on roughly the same plot of land (see Methods). We justify our Theme Two predictions on the grounds that attitudes change slowly in individuals over time but vary widely between individuals in any one period. Moreover, we predict that the deeper the respondent's recall window, the more consistent his or her perceptions across T1 and T2, i.e. we expect that an individual who has spent at least seventeen years on a plot of land will be more likely in T2 to repeat something they reported in T1. We presume a similar trend for a T2 respondent who is closely related to the person interviewed in T1 on the same plot of land. A newcomer's perceptions are less likely to match those of the T1 respondent, especially if s/he never knew the T1 respondent.

Methods

Time 1 (1994)

During March–April 1994, Lisa Naughton (LN) worked with three field assistants (P. Baguma, F. Mugurusi and P. Kagoro, all local residents) to interview all households across six villages within 500 m of the park about their perceived crop losses. Interviews were conducted in local languages (Rutoro or Rukiga) and the same introduction and standard questions were used throughout. All individuals who stated that animals damaged crops on their farms were asked to list the animals visiting their farm[2] and to identify the three worst. The first species mentioned was considered the worst in subsequent analyses. Respondents were also asked what crops were most often damaged, whether the problem was changing over time and what defensive strategies they and their neighbours used. We also asked individuals to comment on the 1993 decision to upgrade Kibale from a reserve to a park and we recorded the respondents' ethnicity and sex. Off-farm employment was recorded if any family member reported earning wages as a labourer during the past five years. A farm's distance to the forest boundary was measured in metres from the point closest to the forest. For full description of methods, see Naughton-Treves (1997, 1998). AT assisted with T1 field data collection, analysis and writing.

Time 2 (2011-2012)

During August 2011, LN returned to Kibale with Jessica L'Roe (JL) and Andrew L'Roe (AL) to set up resampling in the same six villages. To ensure continuity in methods and monitoring sites, two of the original assistants from 1992–1994 (F. Mugurusi and P. Baguma) helped find the prior study's transect lines and train two new assistants (local residents R. Karamagi [RK] and J. Ogwang [JO]). JL, AL, RK and JO collected data for the next twelve months (September 2011 to August 2012). Interviews about human-wildlife interactions and ranking of crop-raiding wildlife followed the 1994 template. Again, we attempted to interview all residents within 400 m of the park as well as T1 individuals who had moved within the study area. In T2, questions were added about the relative severity of crop loss to wildlife versus other problems. Land sales and subdivisions were recorded so as to track field ownership, as were crop field plantings, fallows, tea and eucalyptus plantations (L'Roe 2016 describes trends in land ownership during the study). Sample size varied in both time periods because not all respondents answered all questions.

All respondents' participation was voluntary and anonymous. Human subjects' clearance was obtained in 1992–1994 from the Uganda National Council for Science and Technology, the Ugandan Forest Department, the Ugandan Park Service and the University of Florida, Gainesville. In 2011–2012, clearance was obtained from the Uganda Wildlife Authority, the Uganda National Council for Science and Technology. and the University of Wisconsin–Madison.

Results

Respondents' Attributes and Socioeconomic Changes, 1994 and 2011-2012

Ninety-four citizens were interviewed in six villages bordering Kibale in 1994 (62 per cent men, 66 per cent Toro ethnicity) (figure 7.1). Seventeen years later, 138 citizens were interviewed on roughly the same plots of land in the same villages (60 per cent men, 69 per cent Toro ethnicity). Thirty-four individuals were interviewed in both T1 and T2. Of these repeat respondents, twenty-seven were interviewed in roughly the same plot of land in both periods (SAME-PERSON/SAME-LAND). 'SAME-LAND' refers to properties that overlap to some degree in T1 and T2. Precise land matches were not possible because nearly all property boundaries changed due to land sales and subdivisions during the study. Seven of the thirty-four repeat respondents had shifted to a new plot of land – i.e. their T2 plot of land did not overlap with the T1 plot (SAME-PERSON/DIFFERENT-LAND). All but 1 of the 104 individuals interviewed only in T2 used or owned land that physically overlapped to some extent with a T1 property (DIFFERENT-PERSON/SAME-LAND). Again, precise matches were not possible given that the majority of landholdings changed, often significantly (e.g. a T1 parent subdivided her property among children and sold one plot to a newcomer, or a T1 landowner doubled his farm size by purchasing adjacent plots). Of the 104 DIFFERENT-PERSON/SAME-LAND respondents, 13 were spouses of the T1 landholder, 51 were relatives (children, grandchildren, siblings, nieces, cousins, etc.) and 40 were non-relatives.

Key Land Use and Social Changes at the Park Boundary during Study

The most dramatic change in land use happened at the northernmost village, Sebitoli, a long-standing elephant-raiding 'hotspot' (Game Dept Archives 1951; Naughton-Treves 1997). Here all but two of the seventeen T1 respondents abandoned their land or sold it off to tea companies. In the remaining five villages we observed a dramatic decline in brewing bananas, a cash crop cultivated by 77 per cent of the households (HH) in T1 versus 48 per cent of HH in T2 ($X^2 = 66$, $p < 0.0001$). This change matches a regional decline attributed to banana weevil problems (Komarek and Ahmadi-Esfahani 2011). Land subdivisions were common and farm parcel size shrank by roughly half during the seventeen-year interval (from an average of ~1.4 to ~0.7 ha). Precise measurement of parcel size is difficult due to overlapping claims and complex tenurial arrangements, only some of which are formalized. But apparently while parcel size decreased, by 2012 more households reported owning multiple landholdings (T1: 17 per cent of 94 HH, T2: 44 per cent of 136 HH). Residence time for T1 residents averaged twenty-one years (standard deviation (SD) = 15, range: 2–59). Residents in T1 who self-identified as Toro had longer residence times than did those who identified themselves as Kiga or 'Other' ethnicity (Toro average = 24 years, SD = 16; Kiga average = 18 years, SD = 9.5; 'Other' average = 8.3, SD = 7.4). Most T1 respondents reported that their land was 'mostly fallow' or 'mostly cultivated' when they acquired it (forty-nine [54 per cent] and thirty [33 per cent], respectively, n = 90). In T2 roughly half of the respondents said they were born on the land they now resided on (fifty-seven born on farm, n = 105). Of the fifty-seven born on farm, fifty (88 per cent) self-identified as Toro. Of the forty-eight (42 per cent) who moved to the plot of land or acquired it sometime after they were born, twenty were Toro. Thus as per common understanding, Toro respondents generally had longer residence times in the study area than did Kiga or other ethnic groups (Naughton-Treves 1999).

The average household wealth improved during the study by various measures. The percent of households with durable metal roofing ('Mabati') more than doubled from T1: 40 per cent, n = 91 HH, to T2: 89 per cent, n = 138 HH, $X^2 = 224$, $p < 0.001$). In 1994 only 1 per cent of respondents grew tea, a 'rich man's crop' according to our respondents, whereas by 2012 13 per cent did ($X^2 = 191$, $df = 1$, $p < 0.0001$). In 2012 we counted assets that we did not even think to look for in 1994 (e.g. 20 per cent HH owned motorcycles in T2 and one owned a satellite dish, n = 138). Access to outside employment remained roughly the same (T1: 59 per cent, T2: 65 per cent reported some access to outside employment during past approximately five years), although this result should be interpreted cautiously given variation in how respondents interpreted 'outside employment' and their recall window. The number of respondents employing workers on their land increased (T1: 5 per cent vs. T2: 15 per cent). The average reported number of cattle and goats per household stayed roughly constant.

Crop-Loss Reports and 'Worst Animals' Over Time

In 1994 many respondents expressed concern about Kibale Forest's 'upgrade' to a national park and 60 per cent anticipated increased human-wildlife conflict (n = 89). Roughly half (51 per cent) also feared an increased presence of dangerous animals, particularly lions (named by 74 per cent of the forty-three respondents concerned about dangerous animals), leopards (16 per cent), hyenas (16 per cent) and elephants (12 per cent). Some respondents feared park authorities would evict them from their land (28 per cent, n = 89). A few (7 per cent) thought the new park status would attract more tourists and 18 per cent anticipated no impact from the conservation upgrade (n = 89).

We compared change over time in the patterns of conflict with wildlife and local perceptions in two ways (as per Theme One and Two). First we compared group-level data from each period to describe general patterns in the reported frequency of crop-raiding species, most vulnerable crops and coping mechanisms. We also used the group-level data to test whether basic predictors (e.g. distance from park, village, ethnicity and sex of respondent) showed a significant association with reports of raiding by certain species. Second, we used time-series data to explore change in perceptions at the individual level and paired the data in two ways: (1) SAME-PERSON/SAME-LAND versus DIFFERENT-PERSON/SAME-LAND, and (2) SAME-PERSON/SAME-LAND versus SAME-PERSON/DIFFERENT-LAND.

Group Level Data Analysis

Across both periods, respondents reported a total of eleven wildlife species visiting their farms. A methodological problem prevents us from directly comparing the number of species listed. Namely, in T1 we asked respondents to name all the wild animals visiting their land and produced a tally. In T2, we asked respondents to confirm (Yes/No) whether they had observed eleven different species on their land. Not surprisingly, more species were registered under the T2 survey method (T1 mean = 3.3 species mentioned, SD = 1.9, n = 93; T2 mean = 7.5 confirmed, SD = 2.6, n = 134). In T1 the five species most often reported 'visiting' respondents' farms were: redtail monkeys (71 per cent), bushpigs (60 per cent), baboons (57 per cent), chimpanzees (33 per cent) and elephants (31 per cent) (n = 94 respondents). In T2 138 respondents reported a similar suite of top-five 'visiting' species but in a different order: baboons (99 per cent), elephants (92 per cent), redtail monkeys (84 per cent), cane rats (72 per cent) and bushpigs (71 per cent).

Table 7.1. Individual-, farm- and village-level predictors of reported crop loss to four wildlife species around Kibale National Park, Uganda: 1994 and 2011–2012.

	Reported # wildlife species χ^2	Elephant Visit Y/N χ^2	Elephant Rank χ^2	Baboon Visit Y/N χ^2	Baboon Rank χ^2	Bushpig Visit Y/N χ^2	Bushpig Rank χ^2	Chimpanzee Visit Y/N χ^2	Chimpanzee Rank χ^2
Time One (1994, n = 94 households)									
Whole model[1]	38**	76***	48***	26**	NS	36***	35**	28**	NS
Distance to park (m)	5*	20***	10**	11**	NS	6**	NS	NS	NS
Village	12*	56***	41***	NS	NS	28***	71***	NS	NS
Sex	4*	NS	NS	NS	NS	NS	NS	NS	NS
Time Two (2011–2012, n = 124 households)									
Whole model[1]	15*	28**	NS	NS	NS	18*	NS	28**	NS
Distance to park (m)	F = 7*[2]	14**	5.5*	3.5*	NS	15*	NS	13**	NS
Village	12*	NS	NS	NS	NS	15*	NS	NS	NS
Sex	NS	NS	NS	NS	NS	NS	NS	NS	NS

1. Whole model includes: Farm's distance to park (m); Village where farm located (5 different villages on park edge); Ethnicity of respondent (Toro, Kiga or other); Sex of respondent (M/F). Results for ethnicity are not presented because it was not predictive in univariate tests and it was significantly associated with village.
2. Because of collinearity issue in Time Two, we first took simple linear regression and used the residuals in the GLM.
*** $p \leq 0.0001$, ** $p \leq 0.001$, * $p \leq 0.05$
Visits by redtail monkeys were frequently reported in Time One and Two, but were not predicted by any of the variables (see Results section above).

We then compared the average distance from the park boundary for households reporting crop loss to five species in T1 and T2 (table 7.1). The distance from park increased on average for all species, but was only significant for elephants (T1 average distance from park for elephant complainants = 57 m, SD = 189, n = 29; T2: 156 m, SD = 208, n = 120, t = 2.5, p < 0.017).

Turning to the average ranking of 'worst animal', in T1 bushpigs were ranked worst, followed by baboons, then redtail monkeys and finally elephants (1–4, with 1 being worst, we assigned the lowest rank, 4, to an animal not mentioned): bushpigs mean ranking = 2.7, median = 3, SD = 1.3; baboons mean = 2.9, median = 4, SD = 1.2; redtail monkey mean = 3.0, median = 3.5, SD = 1.2; elephants mean = 3.4, median = 4, SD = 1.1; n = 94). Respondents in T2 ranked baboons worst, followed by elephants and redtail monkeys (baboons mean = 1.8, median = 1.0, SD = 1.0; elephants mean = 1.9, median = 2, SD = 1; redtails mean = 3.4, median = 4, SD = 0.8). In T2 respondents were more concerned about elephants, baboons and redtail monkeys and less about bushpigs. Further analysis indicated that the average concern about elephants rose the most in three villages where elephant raiding was not observed in T1 (Kruskal Wallis X^2 = 16.5, df = 5, p = 0.0055). Other species (e.g. chimpanzees) showed relatively constant rankings in T1 and T2.

The observed changes in worst animal rankings are corroborated by additional T2 data. When asked if they perceived a change in the abundance of each crop-raiding wildlife species, 66 per cent of respondents reported more elephants and 46 per cent perceived an increase in baboons (n = 134). A majority reported no change in raiding by redtail monkeys (66 per cent) and chimpanzees (63 per cent). Meanwhile most respondents believed bushpig numbers had declined (68 per cent, n = 134).

Respondents also ranked crops by vulnerability to crop-raiding. These rankings showed little change (1–4 ranking, 1 being most vulnerable; T1: maize and sweet potato average 2.8 and 2.7 respectively, banana 2.9 and cassava 3.3, n = 92; T2: maize 2.4, sweet potato 2.5, banana 3.0 and cassava 3.2, n = 131).

Next we used a multivariate model to test whether (1) farm distance to park edge, (2) village, (3) ethnicity or (4) sex predicted respondents' reports of species' visits on their farms or ranking of worst animals. We first checked for associations between the four predictor variables and found that village and tribe were significantly associated in both T1 and T2. In T2 only, distance and village were also significantly associated, namely Kabucikire respondents lived an average ~42 m farther from the park edge than other villages' residents, likely due to forest clearing during the study. We did not include village and distance from park edge in the same analysis for T2 data, therefore this does not affect our subsequent analyses.

In T1 the best predictor of the reported total number of wildlife species visiting a farm was distance to park (GLM whole-model test: X^2 = 38, df = 9, p < 0.001, distance to park: X^2 = 5, p = 0.02, sex: X^2 = 4, p = 0.05, village: X^2 = 12, p = 0.03) (table 7.1). We then used the same four predictors to model respondents' reports of each individual wildlife species' visits to their farms in T1. Elephants were more likely to be reported in certain villages (GLM whole model test: X^2 = 76, df = 9, p < 0.0001, village: X^2 = 56, df = 5, p < 0.0001) and by those respondents living closer to the park (distance from forest: $X^2 = 20$, df = 1, p < 0.0001). Bushpigs were also significantly more often reported in certain villages and closer to the park (GLM whole model: $X^2 = 36$, df = 9, p < 0.0001, village: $X^2 = 28$, df = 5, p < 0.0001, distance from park: $X^2 = 6$, df = 1, p = 0.01). For reports about baboons, the whole model test was significant and only distance to park was a significant predictor (GLM whole model: $X^2 = 26$, df = 9, p = 0.002, distance to park: $X^2 = 11$, df = 1, p = 0.0007). Men were more likely than women to report the presence of chimpanzees, but no other variable proved significant for this species in T1 (whole model: $X^2 = 28$, df = 9, p = 0.0008; sex: $X^2 = 5$, df = 1, p = 0.03; 46 per cent of men mentioned chimpanzees, 9 per cent of women mentioned chimpanzees, n = 94). None of the variables we tested predicted reported visits by redtail monkeys and ethnicity did not predict reported visits by animal.

Turning to 'worst raider' rankings (also in T1), for elephants the GLM whole model proved significant ($X^2 = 48$, df = 9, p < 0.0001), village ($X^2 = 41$, df = 5, p < 0.0001) and the distance to the park ($X^2 = 10$, df = 1, p = 0.0002). The whole model also proved significant for bushpig rankings in T1 ($X^2 = 35$, df = 5, p < 0.001) based on the village she or he lived in ($X^2 = 71$, df = 5, p < 0.0001). The whole-model test was not significant for rankings of baboons, redtail monkeys or chimpanzees in T1. In short, T1 respondents in certain villages and close to the park were more likely to report visits by bushpigs and elephants and rank these as worst pests. Respondents from all villages were equally likely to report baboon visits in T1, but those closest to the park were most likely to rank this as a worst animal.

In T2, distance to park remained a strong predictor of the number of species respondents reported visiting their farms (n = 129, F = 7, p = 0.011). Having controlled for distance (due to collinearity in T2 of distance and village), we tested its residuals with the three other predictors. We found village was weakly predictive of total number of species visiting a farm (GLM whole model: $X^2 = 15$, df = 8, p = 0.054, village: $X^2 = 12$, df = 5, p = 0.04). Elephant visits in T2 were predicted only by distance to park (whole model: $X^2 = 28$, df = 9, p = 0.0008, distance: $X^2 = 14$, df = 1, p = 0.0002). A similar result held for bushpigs (whole model: $X^2 = 15$, df = 5, p = 0.011; distance to park: $X^2 = 15$, df = 5, p = 0.022) and chimpanzees (whole model: $X^2 = 28$, df = 9, p = 0.001, distance to park: $X^2 = 13$, df = 1, p = 0.003).

Turning to worst-animal rankings in T2, only distance to park significantly predicted elephant ranking (whole model: $X^2 = 5.5$, df = 1, p = 0.022). For baboon rankings in T2, the whole model was approaching significance so we dropped the weakest predictor and found a trend that people living closer to the park were more likely to rank baboons as worst ($X^2 = 3.5$, p = 0.06). We found no significant predictors for rankings of chimpanzees and redtails in T2. Bushpigs apparently remained more prevalent in certain villages in T2, but even in these villages people no longer ranked them among the worst pests.

Testing Individual Attitude Change Over Time

We next focused on longitudinal change in perceptions of 'worst species' ranking and general trend in perceived severity of crop-raiding. In twenty-seven cases we interviewed the same person on roughly the same plot of land in T1 and T2 (n = 27

SAME-PERSON/SAME-LAND). Due to high turnover in land ownership, far more often we interviewed different individuals in T1 and T2 who we matched if their respective plots of land in T1 and T2 overlapped (n = 115 DIFFERENT-PERSON/SAME-LAND). This method of matching created potential pseudo-replication. Take for example a T1 landowner who subdivided his land, sold parts to three people but kept one piece. In T2 we interviewed the original landowner again as well as the three new owners. This created four pairs of T1–T2 comparisons and in each pair the T1 data were the same (corresponding to the owner's perceptions in T1 for the entire plot). Similarly, although less often, if a T2 respondent bought land from two T1 owners, two pairs were created for analysis, each with the same T2 data.

To compare pest rankings over time, we calculated the absolute difference in T1 and T2 for each species (range 0–3) and also looked at the direction of change –3 to +3. T1 vs. T2 rankings changed more for elephants than for any other raiding species. On average, elephants shifted two steps in the 'worst' pest ranking for both the SAME-PERSON/SAME-LAND and DIFFERENT-PERSON/SAME-LAND matches (mean and mode = $\Delta|2|$). But the magnitude of change in elephant rankings over time was significantly greater for DIFFERENT-PERSON/SAME-LAND matches than SAME-PERSON/SAME-LAND (two sample median test with normal approximation, non-parametric $z = -3.4$, $p = 0.0007$). Whereas 35 per cent of the DIFFERENT-PERSON/SAME-LAND matches shifted three ranks, only 7 per cent of the SAME-PERSON/SAME-LAND matches did so. Twice as many of the SAME-PERSON/SAME-LAND matches kept constant rankings as in the DIFFERENT-PERSON/SAME-LAND matches. In T2, repeat respondents did not rank elephants as high as did the T2 only respondents ($z = -3.36$, $p = 0.0008$).

SAME-PERSON/SAME-LAND matches also showed more consistency in rankings of bushpigs over time (less absolute change, $z = -2.5$, $p = 0.032$). In terms of direction of change, DIFFERENT-PERSON/SAME-LAND matches had a greater change in a positive direction (bushpigs ranked lower in T2, $z = 2.1$, $p = 0.038$). By contrast, baboon-ranking changes tended to be greater for SAME-PERSON/SAME-LAND matches (long-term residents less consistent, $z = 1.8$, $p = 0.07$). There was no significant difference in the absolute change or direction of change in rankings for chimpanzees and redtail monkeys.

We next examined data for repeat respondents only, but comparing SAME-PERSON/SAME-LAND (n = 27, as above) to SAME-PERSON/DIFFERENT-LAND (n = 7). We found no significant difference in the amount or direction of change in ranking elephant, bushpig, redtail monkeys and chimpanzee among worst pests. Only in the case of baboon rankings over time did a significant difference appear; those who stayed on the same plot of land between T1 and T2 were less consistent and more likely to rank baboons worse than those who moved ($z = 2.2$, $p = 0.028$ for absolute difference of ranking, $z = -2.0$, $p = 0.049$ for direction of difference of ranking).

Finally, we tested whether the degree of relatedness between the T1 and T2 respondents correlated with the amount and direction of perception change. We categorized 'relatedness' between T1 and T2 respondents in rank order: SELF = 1 (n = 34), SPOUSE = 2 (n = 23), RELATIVE = 3 (n = 50; including children, cousins, siblings, nieces, nephews, aunts, uncles, grandchildren), and NON-RELATIVE = 4 (n = 58). Relatedness proved a significant predictor of consistency in ranking for elephants and a tendency for bushpigs (elephants: X^2 = 17, df = 3, p = 0.0007, bushpigs X^2 = 7, df = 3, p = 0.07). In both cases, SELF matches were most consistent, then spouse, then family and non-related matches least consistent. For complaints about redtail monkeys, baboons and chimpanzees, respondents' relatedness and their changes in perception were unrelated. To help interpret this result we tested relatedness versus the two fundamental factors shaping crop-raiding patterns: village and distance from park. We found no association between village and relatedness. But an interesting pattern surfaced in the distance to park test. T2 repeat respondents (i.e. SELF matches) tended to reside farther away from the park in T2 than the non-SELF matches (Welch test t = 2.1, p = 0.044, SELF match distance in T2 mean = 267 m, SD = 341 vs. non-SELF distance in T2 mean = 143 m, SD = 125). This suggests a higher turnover rate of land at the forest edge, i.e. long-term residents apparently were more likely to sell off or bequeath plots of land closer to the park than those farther from the park. A closer look at relatedness versus distance shows that the level of relatedness significantly explained distance to the park in T2 (Welch test F = 4.1, p = 0.013, SELF match distance in T2 mean = 267 m, SD = 341, n = 34; SPOUSE match T2 distance mean = 190, SD = 199; RELATIVE match T2 distance mean = 167 m, SD = 114, n = 50; NON-RELATIVE match T2 distance mean = 111 m, SD = 106, n = 58). The high standard deviations (especially SD = 341 m for SELF matches) signals marked variability. Cautious interpretation is also warranted given the pseudo-replication issue.

Coping Strategies in T1 and T2

The most commonly reported response to the threat of crop loss to wildlife was guarding (T1 = 81 per cent, n = 90, T2 = 93 per cent, n = 117). In both time periods less than 5 per cent of respondents reported building trenches or fences as barriers to wildlife. Very few (less than 5 per cent) reported hunting or trapping, although some T2 respondents mentioned that official 'baboon' hunters had visited their village. T2 respondents were more likely to report that the presence of wildlife prevented them from planting certain crops (T1: 33 per cent, T2: 64 per cent) or cultivating part of their farm (T1: 24 per cent, T2: 44 per cent, nearly always the land adjacent to the park). The majority (76 per cent) of T2 respondents explicitly stated that they changed their cultivation practices due to wildlife crop-raiding (n = 133). Of these individuals, 45 per cent reported changing to food crops less palatable to wildlife (e.g. yams) or crops with shorter planting seasons (e.g. groundnuts). Many switched from growing crops to planting eucalyptus trees or tea on at least part of their farm (33 per cent and 14 per cent respectively) and a few either left land in fallow (6 per cent) or sold their land (6 per cent) (n = 71, multiple responses allowed). In Sebitoli, the village of greatest elephant losses, all but two of the seventeen T1 respondents sold their land to the local tea company or left their land fallow. Sebitoli residents' struggle with elephants echoes colonial accounts from certain parts of the Toro region where 'the few remaining inhabitants were no longer able to repel . . . the depredations of game' (Osmaston 1959). A district warden visiting Toro in 1916 described 'wanton destruction' and 'complete devastation' of farms by elephants such that 'the [human] populace has been forced to move elsewhere' (Game Dept Archives 1924).

Discussion

By comparing our respondents' reports, we infer that wild animals ventured farther from the park edge to raid crops in 2011–2012 than in 1994. The high variability in the data and modest scale of increase (tens of metres further) warrants cautious interpretation, but in the case of elephants, the expanded distance of complaint proved significant (table 7.1). Not only has elephants' reported average foray distance from the park more than doubled, elephants are now raiding crops in villages that were free from such problems in T1. Given Tropical Africa's dire loss of elephants due to poaching in recent years, an elephant population growth problem seems incongruous, but this might be the case at Kibale. The fact that 'village' was no longer a predictor in T2 is a sign that elephant raiding is extending beyond hotspots, not that hotspots are disappearing. In 2011 Catherine Mackenzie surveyed a subsample of thirty-five farms from the six villages studied in 1992–1994 and she estimated that elephant raiding had increased significantly (66 per cent more farms reported elephant damage than in the first study). Apparently, as respondents feared in T1, elephant raiding has worsened in the seventeen years following Kibale's upgrade to national park status (although problems with lions and hyenas did not increase as feared). Counter to expectation, the near doubling of the human population at Kibale's edge did not reduce reported raiding distances for elephants nor other wildlife (Newmark 1996). The underlying causality of elephant raiding expansion is unclear, especially without systematic data on Kibale's elephant populations. Evidence from southern Kenya suggests that food crops are particularly attractive to elephants because they are easier to digest, high in energy and low in secondary compounds, which enables male elephants to reach larger body size and achieve greater mating success (Chiyo et al. 2011). Therefore, reports of increasing crop-raiding may reflect the higher elephant populations estimated in Kibale (Wanyama et al. 2010). Alternately, elephants at Kibale may now be more habituated to people, a dangerous trend, if indeed this is happening. Raiding distances have apparently expanded for elephants (and possibly other species), and on occasion, raiding happens at considerably greater distances than those we measured (Mackenzie 2012). Ultimately, however, the zone of frequent crop loss is largely confined to the first 200 m from the park edge. Similarly, Patrick Chiyo (2000) found elephant damage generally confined to within 200 m of Kibale's boundary. This pattern concurs with local residents' advice (and those neighbouring other East African parks) that the best defence against elephants and other large game is to have an active farm between you and the park (Newmark 1996; Hill 1997).

The narrowness of the risk zone is good news for farmers in the regions, but it also signals that Kibale's wildlife remain insularized (Hartter and Southworth 2009). During our study, forest was generally maintained in the park, but beyond park boundaries tall canopy forest declined by approximately 50 per cent (Naughton-Treves, Alix-Garcia and Chapman 2011). Even among the adaptable 'raiding' mammal species we studied, only redtail monkeys seem able to persist outside the park in the densely settled agrolandscape (Chapman et al. 2010). Our study also revealed intriguing social outcomes, e.g. a higher turnover rate in land ownership at the forest edge, declining size of land plots with time and an increase in smallholder tea plots. These outcomes deserve more attention (L'Roe 2016) particularly given conservationists' concern about population growth around park edges (Wittemyer et al. 2008).

Bushpigs offer an intriguing counter to the familiar story of expanded raiding and growing local concern about crop loss to wildlife. Bushpigs fell from being worst pest in T1 to seldom even ranked during our study. Again, causal explanation is elusive without systematic data on bushpig numbers. The decline in complaints about this species may have something to do with the Ugandan Wildlife Authority's decision in 2000 to legalize supervised hunting of three species of declared vermin: baboons, bushpigs and vervet monkeys (Olupot, McNeilage and Plumptre 2009). On the other hand, in T2, citizens neighbouring Kibale only reported visits from official 'baboon' hunters, not bushpig hunters. The fact that baboons did not enjoy similar increase in tolerance may be because Toro citizens generally consider it taboo to eat primates. However, many enjoy eating bushpigs. As one respondent told us in Rurama, 'bushpigs were a problem but we ate them'. Elsewhere in Uganda, hunters reported greater motivation to kill bushpigs for 'products' (e.g. bushmeat) than for crop protection (Olupot, Barigyira and Chapman 2009). The reverse was true for baboons (ibid.). Even in areas where there is no taboo on eating primates, baboon or vervet monkey meat costs less per kg than bushpig meat (ibid.). Ultimately we are unable to conclude whether declining complaints about bushpigs signals improved tolerance for this species or that bushpig numbers have declined.

Consistency of Attitudes

Although we measured attitudes in a highly simplified manner (e.g. worst pest ranking), our results offer some insight about change in attitudes to wildlife over time. As predicted, when we interviewed the same person in T1 and T2, we found less variability in worst-animal rankings and crop-loss complaint than when we interviewed different people (even controlling for the same location, sex and ethnicity). The fact that a simple measure of 'relatedness' proved significant in predicting variability is intriguing. Greater consistency in crop-raiding complaints and worst-animal rankings among family members than among unrelated respondents may be due to deeper recall windows about crop loss to wildlife over previous years. Relatedness between respondents may also be a proxy for more closely shared value orientations (Teel and Manfredo 2010), although it is not ultimately clear how value orientations shape worst-animal rankings.

Methodological Challenges

Too often crop-raiding by wildlife is assessed only at hotspots and only via public meetings. If such results from a hotspot are then extrapolated to an entire region, it can make a volatile issue worse, potentially exacerbating local resentment of a park and/or masking elevated vulnerability of a narrower group of citizens. We are generally confident that our respondents answered interview questions as best they could due to our long-term engagement in the community. The fact that in both periods, we regularly canvassed crop losses while conducting interviews may have signalled our interest in accuracy. During pilot research in 1992, before any kind of field measurements started or relationships were established, a few respondents reported that crop-raiding elephants were causing people to starve for kilometres beyond the park edge. Such exaggerated complaints became fewer the longer we worked in the region. Ideally, researchers should compare perceptions of loss and worst-pest rankings to systematic field measurements of loss. Disparities and congruence in the two forms of data can be revealing (Naughton-Treves 1997).

With regard to more specific methods, it proved unexpectedly challenging to match 'farms' over time due to widespread land transactions. In a few cases, we were not sure what specific location a T2 respondent was referring to during the interview, a problem compounded by the advanced age of some respondents. Two interviews were dropped for this reason. Improvements in GPS accuracy have already solved some of the technical problems we faced with our 1994 data, but the complexity of land tenure and affiliated social relations make 'controlling' for location difficult. However, given that crop loss to wildlife is usually patchy in distribution and hotspots are common, spatially explicit data are important for both cross-sectional and longitudinal study of attitude change.

Conclusion

Our interview data from 1994 and 2011–2012 suggest that some of Kibale's wildlife species now raid crops at modestly greater distances from the park edge. In the case of elephants, this expansion has been significant and the risk of elephant raids is no longer confined to a few 'hotspot' villages. Local residents are concerned about this growing risk and are now more likely than seventeen years ago to alter their planting to reduce crop losses to wildlife. The one species to become better tolerated by residents is the bushpig, an animal the Uganda Wildlife Authority designated as vermin and permissible to kill in supervised hunts beginning in 2000. Despite the same status change, baboons remain highly unpopular among farmers at Kibale's edge.

Our exploration of change in individual attitudes over the seventeen-year period reveals that reports of wildlife presence and rankings of worst animal are more likely to match if the same individual respondent is interviewed in T1 and T2. Similarly, we found greater similarity in perceptions when in T2 we interviewed spouses of T1 respondents. Other family members showed greater variation in perceptions followed by non-relatives. The causal explanation for this pattern of more highly congruent perceptions with closer relations is not entirely clear. Deeper recall window and greater familiarity with raiding history for a given site might have an effect, and/or closer relatives are more likely to share value orientations.

Lisa Naughton-Treves is Professor of Geography at the University of Wisconsin–Madison. Her research concerns the social dimensions of biodiversity conservation, with particular emphasis on protected areas and land-use conflicts in the tropics. She has long-term field studies in Uganda, Ecuador and Peru. She directed UW–Madison's Land Tenure Center, chaired the graduate program in Conservation Biology and Sustainable Development and now chairs the Geography Department.

Jessica L'Roe studies changes in land use in the tropics. She focuses especially on agriculture-forest dynamics in the Amazon and East Africa under changing economic and policy conditions. She has a PhD in Geography and a MA in Agricultural Economics from the University of Wisconsin–Madison and has recently joined the faculty at Middlebury College.

Andrew L'Roe earned his PhD in the Department of Forest and Wildlife Ecology at the University of Wisconsin–Madison. His research is focused on changes in the ownership, management and use of forestland. He studies the effects of public policy tools like tax programs, conservation easements and forest certification on the conservation of privately owned forests

Adrian Treves earned his PhD at Harvard University in 1997 and is now an Associate Professor of Environmental Studies at the University of Wisconsin–Madison. His research focuses on ecology, law, the public trust and agroecosystems where crop and livestock production overlap carnivore habitat. He and his students work to understand and manage the balance between human needs and carnivore conservation. He has authored more than a hundred scientific papers on predator-prey ecology or conservation. Most recently Dr. Treves has been writing and speaking on the public trust doctrine.

Notes

1. Farm-size and population-density figures refer only to smallholdings. More extensive land uses within 1.5 km of Kibale's boundary include forest fragments, tea estates and grassland (Mugisha 1994; Hartter, Goldman and Southworth 2011).
2. The Rutoro word for animals, *ebisoro*, does not include birds. Although several respondents reported crop loss to birds, the frequency with which people complained about birds may be underestimated.

References

Aluma, J., et al. 1989. *Settlement in Forest Reserves, Game Reserves, and National Parks in Uganda*. Kampala, Uganda: Makerere Institute of Social Research, Makerere University.

Bruskotter, J.T., R.H. Schmidt and T.L. Teel. 2007. 'Are Attitudes toward Wolves Changing? A Case Study in Utah', *Biological Conservation* 139: 211–218.

Chapman, C.A., et al. 2010. 'Understanding Long-Term Primate Community Dynamics: Implications of Forest Change,' *Ecological Applications* 20: 179–191.

Chiyo, P.I. 2000. 'Elephant ecology and crop depredation in Kibale National Park, Uganda.' Senior thesis, Makerere University, Kampala, Uganda.

Chiyo, P.I., et al. 2005. 'Temporal Patterns of Crop Raiding by Elephants: A Response to Changes in Forage Quality or Crop Availability?' *African Journal of Ecology* 43: 48–55.

———. 2011. 'No Risk, No Gain: Effects of Crop Raiding and Genetic Diversity on Body Size in Male Elephants', *Behavioural Ecology* 22: 552–558.

Dickman, A.J. 2010. 'Complexities of Conflict: The Importance of Considering Social Factors for Effectively Resolving Human–Wildlife Conflict', *Animal Conservation* 13: 458–466.

Ericsson, G., and T.A. Heberlein. 2003. 'Attitudes of Hunters, Locals, and the General Public in Sweden Now that the Wolves are Back', *Biological Conservation* 111: 149–159.

Gachago, S., and J. Waithaka. 1995. *Human–Elephant Conflict in Kiambu, Murang'a, Kirinyaga, Embu and Meru Districts*. Nairobi: Kenya Wildlife Service.

Game Dept Archives. 1924. *Ugandan Game Department Archives*. Entebbe: Government of Uganda.

———. 1951. *Ugandan Game Department Archives*. Entebbe: Government of Uganda.

Hartter, J. 2007. 'Landscape Change around Kibale National Park, Uganda: Impacts on Land Cover, Land Use, and Livelihoods.' PhD Dissertation. Gainsville: University of Florida.

Hartter, J., A. Goldman and J. Southworth. 2011. 'Responses by Households to Resource Scarcity and Human–Wildlife Conflict: Issues of Fortress Conservation and the Surrounding Agricultural Landscape', *Journal for Nature Conservation* 19(2): 79–86.

Hartter, J., and J. Southworth. 2009. 'Dwindling Resources and Fragmentation of Landscapes around Parks: Wetlands and Forest Patches around Kibale National Park, Uganda', *Landscape Ecology* 24: 643–656.

Heberlein, T. 2012. *Navigating Environmental Attitudes*. New York: Oxford University Press.

Henderson, D.W., et al. 2000. 'Human Perceptions before and after a 50% Reduction in an Urban Deer Herd's Density', *Wildlife Society Bulletin* 28: 911–918.

Hill, C.M. 1997. 'Crop-Raiding by Wild Vertebrates: The Farmer's Perspective in an Agricultural Community in Western Uganda', *International Journal of Pest Management* 43: 77–84.

Houston, M., J.T. Bruskotter and D.P. Fan. 2010. 'Attitudes toward Wolves in the United States and Canada: A Content Analysis of the Print News Media, 1999–2008', *Human Dimensions of Wildlife* 15: 389–403.

Kellert, S.R. 1999. 'The Public and the Wolf in Minnesota.' Report for the International Wolf Center, Ely, Minnesota.

Komarek, A.M., and S.Z. Ahmadi-Esfahani. 2011. 'Impacts of Price and Productivity Changes on Banana-Growing Households in Uganda', *Agricultural Economics* 42: 141–151.

L'Roe, J. 2016. 'Land Investment and Land Access Trends among Smallholders near Tropical Forests: Implications for Conservation and Development.' PhD Dissertation, University of Wisconsin–Madison.

Mackenzie, C. 2012. 'Accruing Benefit or Loss from a Protected Area: Location Matters', *Ecological Economics* 76: 119–129.

Mackenzie, C., and P. Ahobyona. 2012. 'Elephants in the Garden: Financial and Social Costs of Crop Raiding', *Ecological Economics* 75: 72–82.

Majić, A., et al. 2011. 'Dynamics of Public Attitudes toward Bears and the Role of Bear Hunting in Croatia', *Biological Conservation* 144: 3018–3027.

Manfredo, M. 2008. *Who Cares about Wildlife?* New York: Springer.
Mugisha, S. 1994. 'Land Cover/Use around Kibale National Park.' Unpublished report by Makerere Institute of the Environment and Natural Resources, Makerere University, Kampala, Uganda.
Naughton-Treves, L. 1997. 'Farming the Forest Edge: Vulnerable Places and People around Kibale National Park, Uganda', *The Geographical Review* 87: 27–47.
———. 1998. 'Predicting Patterns of Crop Damage by Wildlife around Kibale National Park, Uganda', *Conservation Biology* 12: 156–168.
———. 1999. 'Whose Animals? A History of Property Rights to Wildlife in Toro, Western Uganda', *Land Degradation and Development* 10: 311–328.
Naughton-Treves, L., J. Alix-Garcia and C. Chapman. 2011. 'Lessons about Parks and Poverty from a Decade of Forest Loss and Economic Growth around Kibale National Park, Uganda', *Proceedings of National Academy of Sciences* 108: 13919–13924.
NEMA National Environment Management Authority. 1997. *Kabarole District Environment Profile*. Kampala: Government of Uganda.
Newmark, W.D. 1996. 'Insularization of Tanzanian Parks and the Local Extinction of Large Mammals', *Conservation Biology* 10: 1549–1556.
Nyhus, P., R. Tilson and Sumianto. 2000. 'Crop-Raiding Elephants and Conservation Implications at Way Kambas National Park, Sumatra, Indonesia', *Oryx* 34: 262–274.
Ogra, M.V. 2008. 'Human–Wildlife Conflict and Gender in Protected Area Borderlands: A Case Study of Costs, Perceptions, and Vulnerabilities from Uttarakhand (Uttaranchal), India', *Geoforum* 39: 1408–1422.
Olupot, W.M., R. Barigyira and C.A.Chapman, 2009. 'The status of anthropogenic threat at the people-park interface of Bwindi Impenetrable National Park, Uganda', *Environmental Conservation* 36: 41–50.
Olupot, W.M., A.J. McNeilage and A. J. Plumptre. 2009. 'An Analysis of Socioeconomics of Bushmeat Hunting at Major Hunting sites in Uganda', *Wildlife Conservation Society Working Papers*. Bronx, NY: Wildlife Conservation Society.
Osmaston, H.A. 1959. *Working Plan for the Kibale and Itwara Forests*. Entebbe: Government of Uganda Printer.
Plumptre, A., et al. 2001. 'Chimpanzee and Large Mammal Survey of Budongo Forest Reserve and the Kibale National Park.' Unpublished report to the Wildlife Conservation Society, New York.
Struhsaker, T. 1997. 'Ecology of an African Rain Forest: Logging in Kibale and the Conflict between Conservation and Exploitation.' Gainesville: University Press of Florida.
Teel, T.L., and M.J. Manfredo. 2010. 'Understanding the Diversity of Public Interests in Wildlife Conservation', *Conservation Biology* 24: 128–139.
Teel, T., M.J. Manfredo and H. Stinchfield 2007. 'The Need and Theoretical Basis for Exploring Wildlife Value Orientations Cross-Culturally', *Human Dimensions of Wildlife* 12: 297–305.
Treves, A., L. Naughton-Treves and V. Shelley. 2013. 'Longitudinal Analysis of Attitudes Toward Wolves', *Conservation Biology* 27: 315–323.
Turyahikayo-Rugyema, B. 1974. 'The History of the Bakiga in Southwestern Uganda and Northern Rwanda, ca. 1500–1930.' Dissertation, University of Michigan, Ann Arbor.
Wanyama, F., et al. 2010. 'Censusing Large Mammals in Kibale National Park: Evaluation of the Intensity of Sampling Required to Determine Change', *African Journal of Ecology* 48: 953–961.

Wittemyer, G., et al. 2008. 'Accelerated Human Population Growth at Protected Area Edges', *Science* 321: 123–126.

Zimmermann, B., P. Wabakken and M. Dötterer. 2001. 'Human–Carnivore Interactions in Norway: How Does the Re-appearance of Large Carnivores Affect People's Attitudes and Levels of Fear?' *Forest Snow and Landscape Research* 76: 137–153.

8
Conservation Conflict Transformation
Addressing the Missing Link in Wildlife Conservation

Francine Madden and Brian McQuinn

Human-wildlife conflict is widely recognized as a growing threat to many species of wildlife around the world. While it has long been considered a conflict between people and wildlife, in reality it is equally, if not more so, a conflict between people *about* wildlife (Madden 2004; Dickman 2010; Redpath et al. 2013; Madden and McQuinn 2014). This continued misperception is understandable, in part due to the terminology we have become accustomed to using ('human-wildlife conflict'). Moreover, the sight of a partially eaten cow carcass, or the poached body parts of a tiger, provide concrete focal points and act as symbols over which resentment or hostility brew. In reality, there are usually unmet social and psychological needs simmering beneath the surface of the conflict. For this reason, HWC often serves as a lightning rod for channelling, expressing and deepening discontent among diverse stakeholders at all levels.

Indeed, human-wildlife and other conservation conflicts often serve as proxies for underlying social and cultural conflicts, including struggles over group recognition, social, political and cultural identity, as well as struggles to gain power, voice and meaningful participation in decisions that affect not just the wildlife but the people who live with, impact or are impacted by them. Thus, stakeholders in wildlife management and conservation are continually impacted by human conflict (Nie 2003; Madden 2004; McShane et al. 2011; Redpath et al. 2013; Madden and McQuinn 2014).

Unfortunately, conservation efforts are considerably less effective when they fail to account for the often invisible but powerful forces at play between these multiple levels of conflict and the complex web of entrenched interests and identities influencing conservation interventions (Nie 2003; Madden 2004; Madden and McQuinn 2014). Conservation efforts are often focused too narrowly, operating under the assumption that if one addresses the physical, economic and biological components of the conflicts they will create the conditions necessary for coexistence (Breitenmoser et al. 2005; Nyhus et al. 2005; Woodroffe, Thirgood and Rabinowitz 2005; King, Douglas-Hamilton and Vollrath 2011; Packer et al. 2013). Yet these underlying assumptions are incomplete at best, even where some efforts toward stakeholder engagement and conflict resolution are made (Lederach, Neufeldt and Culbertson 2007). We argue that long-term conservation success, which requires community and stakeholder support and commitment, necessitates that these measures be complemented by analysis and actions that enable practitioners to effectively engage the more elusive and less clearly articulated conflict dynamics motivating these disputes (Madden 2004; HWCC 2008; Dickman 2010; Madden and McQuinn 2014). This requires a fundamentally new approach in how conservationists understand and address conflict (Madden and McQuinn 2014). Adapting principles and processes from the peace-building field, conservation conflict transformation (CCT) offers a more appropriate and effective approach for reconciling the deep-rooted social conflicts that typically underpin conflicts over wildlife conservation (Madden and McQuinn 2014). Doing so builds greater receptivity toward and broader ownership of conservation initiatives (Smith and Torppa 2010). This is critical because the social carrying capacity for wildlife – a measure of how much wildlife people are willing to tolerate in or near human-dominated landscapes – for many species will be determined by our successes or failures in this regard (Madden and McQuinn 2014).

In this chapter, we discuss some of the limitations of current conservation practice and outline the need for a CCT approach. We summarize two models for deepening the reader's understanding of conflict in the conservation context and provide a series of case studies that illustrate the potential benefits for broader implementation of CCT in conservation practice. We conclude with a set of recommendations for the conservation field.

Background to CCT Integration in Conservation

In 2003, a diverse group of conservationists from around the world convened a workshop at the 5th World Parks Congress (WPC) on the growing threat of human-wildlife conflict to many species of wildlife. The group recognized that despite differences in geography, culture or species, there were discernible patterns in HWC across these contexts. The group recommended a definition for HWC that the WPC later adopted: 'Human-wildlife conflict occurs when the needs and behaviour of wildlife impact negatively on the goals of humans or when the goals of humans negatively impact the needs of wildlife. These conflicts may result when wildlife damage crops, injure or kill domestic animals, threaten or kill people' (IUCN 2003). Yet, even at that early stage, several WPC participants emphasized that human conflict typically underpins HWC incidents, driven by issues of group ambitions, values, culture, control and worldview that go well beyond material, economic or resource needs (Madden 2004).

In order to advance and expand on the WPC recommendations, several participants assembled from almost forty interested organizations three years later to form the Human-Wildlife Conflict Collaboration (HWCC).[1] Further refining the definition of HWC adopted above was an important first step, and our experience since has highlighted this original definition's shortcomings and need for expansion. Primarily, as mentioned above, the original definition ignored how conservation conflicts often serve as proxies for underlying social and cultural conflicts, including struggles over group recognition as well as social, political and cultural identity. It also failed to account fully for stakeholders' sense of disempowerment and disenfranchisement and impacts on wildlife or the people affected by it. Since 2006, the collaboration has amassed global and multidisciplinary experience in HWC, drawing upon other disciplines, like peace-building and social-psychology, to broaden our understanding of HWC and pioneer a new approach to HWC: conservation conflict transformation (CCT).

CCT recognizes that current strategies to address HWC (including legal and policy determinants that define protection of and intervention regarding wildlife; tactical approaches to protect livestock, crops, personhood and property; financial compensation to offset costs or incentives to encourage physical coexistence; and the use of spatial barriers and zones to delineate and enforce separation of wildlife and people) are not sufficient solutions by themselves. It also suggests reasons why current participatory or multi-stakeholder efforts to resolve conflict and secure a sense of joint ownership of wildlife and HWC solutions fall short (Lederach, Neufeldt and Culbertson 2007).

Integrating an approach in conservation that reduces the social conflict around wildlife, addresses human social and psychological needs, and builds a sense of receptivity and ownership of wildlife conservation initiatives by the people most affected by it is important because together they may impact the resulting social carrying capacity for many species already threatened by other factors, such as habitat loss and climate change. A further exploration of the limits of current practice below are followed by two CCT analytical models; four case studies illustrate how CCT can transform underlying conflicts from an 'us versus them' fight to a more productive starting point for sustainable conservation.

Limitations of Current Conservation Practice

The current focus of conservation efforts on the tangible material and economic costs associated with HWC relies, implicitly, on Abraham Maslow's 'hierarchy of needs' (1954). Maslow's theory hypothesizes that the most fundamental human needs are physiological (food, water, shelter, sleep) and security-oriented (physical, employment, health, property). Maslow further posits that until these basic needs are met, people do not seek out – and are altogether less concerned with – the 'higher-level' social and psychological needs.

Despite its popularity and folk psychology appeal, Maslow's framework has been repeatedly disproven (Coate and Rosati 1988; Max-Neef, Elizalde and Hopenhayn 1989; Burton 1990). In reality, his critics assert, humans pursue the entire suite of human needs simultaneously and doggedly (Marker 2003). Beyond our physiological needs and our desire for security, the social and psychological human needs include, but are not limited to: identifying of self or group in relation to the rest of the world; belonging, love and acceptance; respect and recognition; freedom of choice in decisions and mobility; meaningful action and engagement; and an individual's ability to develop potential capacities and knowledge (Burton 1990; Marker 2003). While these social and psychological needs are often not well articulated, their influence in conflict is unmistakable. According to John Burton, 'when the non-material identity needs of a people are threatened, they will fight' (1984: 121). The lack of recognition or attention to these needs by the conservation field historically helps explain the persistence of what would otherwise seem to be 'irrational' or self-sabotaging responses to attempts to resolve material disputes. The practical result for conservationists is this: if these unmet social and psychological needs are not addressed, conflict persists. Consequently, any effort to address the problem will be, at best, temporary. The CCT approach contends that if the underlying social conflict dynamics are understood and transformed (as opposed to being solved) the solutions that arise are more durable and likely to succeed.

Unfortunately, current conservation practice relies on a set of implicit and largely untested assumptions about what is necessary to achieve human-wildlife coexistence. With some exceptions, assumptions centre on addressing the physical, economic and material needs of people and wildlife. In other words, the assumption that preventing economic or physical loss, or compensating for it, will suddenly translate into support for conservation ignores the fact that removing a negative does not spontaneously create a positive. It further ignores the fact that some of the opposition to conservation efforts stems from *how* conservation efforts are conceived, imposed and implemented. To that end, efforts toward stakeholder engagement reveal another erroneous assumption: that bringing all the stakeholders together for a big meeting or public comment period constitutes effective stakeholder engagement, which is a predominant assumption and status quo approach in conservation (Pidot 2008; Reed 2008; Decker, Riley and Siemer 2012a; Redpath et al. 2013). In fact, these types of efforts can backfire since they give only lip service to the needs for genuine empowerment, voice and recognition that are essential for reconciling deep-rooted conflict (Lederach, Neufeldt and Culbertson 2007; Reed 2008). These assumptions also ignore or only give limited attention to the reality that a lack of empowerment, recognition and control over decisions impacts the psychological receptivity of people living with wildlife (Lederach 2003). So, while these assumptions often include the understanding that some stakeholder involvement is necessary, that understanding fails to recognize that where there is deeper-rooted conflict, it needs to be reconciled before efforts to address the more material issues can truly be settled. Finally, in assessing public attitudes and acceptability toward management decisions, there is an assumption that these attitudes are fixed and based solely on how people value wildlife or proposed wildlife-management interventions (Manfredo et al. 1998, 1999; Manfredo, Teel and Bright 2003; Manfredo and Dayer 2004). In reality, the limits of what is 'acceptable' could be broadened, sometimes considerably, if other human identity needs are met. This is because such snap-shot assessments may be based in part on the current nature of the relationship among parties and/or the current processes used to come to decisions. In other words, this assumption is incomplete because it is based on a failure to recognize the full spectrum of human needs that ultimately influence the highly elastic nature of a community's social carrying capacity for wildlife. So, while wildlife's capacity to adapt to change is limited, if the social and psychological needs of people are met through more effective stakeholder engagement (that targets the deeper-rooted human identity needs) and relationship-building, then evidence suggests human receptivity to creating the conditions for coexistence expands considerably (Lederach 1997; Madden 2004; Lederach 2005; Madden 2006; Frahm and Brown 2007; Smith and Torppa 2010; Madden and McQuinn 2014). The reverse is also true: if we do not address these needs, social carrying capacity for wildlife decreases, sometimes precipitously.

Understandably, there is a growing anxiety in the conservation field that we are running out of time and viable options to save some of the most charismatic species on Earth. A recent publication suggests widespread fencing around protected areas as a solution to the physical conflict with large carnivores (Packer et al. 2013). Regardless of any potential merits of the solution, how this is done would have a dramatic impact on the results. For instance, if communities feel such a 'solution' is being imposed on them, very likely they will ultimately reject it. Indeed, a failure to address the social and psychological needs of people may result in a solution being implemented that will have little real effect on their perception of the conflict and consequent tolerance toward wildlife. Numerous past efforts to erect either electric or chilli pepper fences to prevent elephants from damaging crops have failed. They were either not maintained or implemented, or if implemented they were torn down and in some cases used for illegal snaring (Sitati and Walpole 2006; Ofithile pers. comm. 2012; Mupunga pers. comm. 2013). These fences did not fail because the solution was not viable, but because the means to arrive at the solution lacked genuine community buy-in and because the social conflict between the local community and the conservation authority was not reconciled. As is proposed in the article by Madden and McQuinn (2014), specific interventions may be, in theory, efficient solutions for reducing HWC losses, but if they do not account for the social conflict dynamics they are at higher risk for failure. For instance, one might predict that the presenting conflict will shift to a debate over where the fence was placed with local people arguing that the positioning of the fence unjustly favours the survival of wildlife at the expense of human needs. Or, the fences may be torn down, in which case we may be left in a worse position than before with even fewer emergency tools at our disposal.

For this reason, we argue that long-term conservation success necessitates that measures to address physical, economic and biological components of HWC be complemented by actions addressing the more elusive and less clearly articulated conflict dynamics motivating these disputes (Madden 2004; HWCC 2008; Dickman 2010; Madden and McQuinn 2014). This requires a fundamentally new approach in how conservationists understand and address conflict (Madden and McQuinn 2014). Adapting principles and processes from the peace-building field to the unique realities of conservation, CCT reconciles these deeper-rooted conflicts, creating greater collective motivation and creativity aimed at sustainable coexistence (HWCC 2008; Bay Area Cougar Action Team pers. comm. 2013; Beggs pers. comm. 2013).

An Introduction to Conservation Conflict Transformation

Here we summarize two models that underpin the CCT approach. The first, 'Levels of Conflict', categorizes the different types of conflict that may exist in any conservation context (Madden and McQuinn 2014). This framework provides the practitioner with a means to better understand what is at stake as well as a basis for designing a plan for conflict transformation. Explicit analysis is essential to determining the factors within a system that are for or against coexistence, minimizing the risk of collateral impacts that may derail coexistence efforts, and identifying potential leverage points within the system of conflict (Anderson and Olson 2003; Richardson et al. 2009).

Second, the 'Conflict Intervention Triangle' illustrates the intervention points, which if not adequately considered will become a source for conflict (Madden and McQuinn 2014). While these two models are not intended to offer a comprehensive CCT analytical approach, and additional tools would be needed to understand fully the complex dynamics of conflict, they do offer the practitioner an orientation to CCT. These models further provide a window into what is involved in both the analysis of a conflict and the conceptual design of measures to address the conflict as part of a conservation intervention.

Levels of Conflict: A Model for Analysis

The Levels of Conflict framework enables analysis of the three levels of conflict that may exist in any conflict context: dispute, underlying and identity-based conflict (CICR 2000; Madden and McQuinn 2014). Figure 8.1 shows the levels of conflict.

The first level of conflict, the dispute, is the obvious, physical, easily articulated manifestation of the conflict. In the wildlife conservation context, the traditional definition of HWC presented in the introduction to this chapter essentially encapsulates the dispute level of HWC. For example, this might be the loss of livestock that harms a rancher or the poaching of a rhinoceros that impacts the goals of conservation.

While conflict can exist solely at this level, it rarely does. Rather, the dispute is typically the tip of the conflict iceberg. If conservation programmes focus solely at this level, they might actually exacerbate the conflict.

Underlying conflict, the second level, develops out of a history of unresolved disputes. When there is a history of incidents, disputes or interactions between the parties that were not previously satisfactorily settled, that history gives added weight

Levels of Conflict

```
         /\
        /  \
       /Dispute\
      /--------\
     / Underlying \
    /--------------\
   / Identity-based  \
  /  / Deep-rooted    \
 /_____ \
```

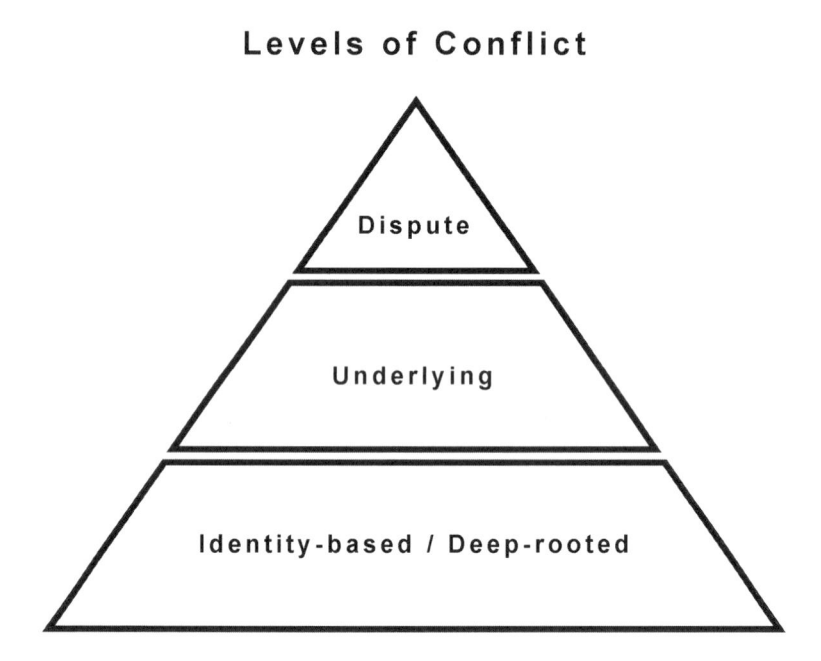

Figure 8.1. 'Levels of Conflict' indicating the three levels of conflict that may exist in any conflict situation. *Source*: Adapted from Madden and McQuinn (2014) and the Canadian Institute for Conflict Resolution (CICR 2002: 73).

to the current dispute. The significance of this history may be ignored or underappreciated by one of the parties who choose to focus solely on the tangible problem. Similarly, the underlying conflict may not be made explicit by an injured party because it is easier to focus one's upset on the current, tangible event, rather than attempt to articulate a past injustice, disrespect or other historical grievance. Examples of history giving rise to underlying conflict might include: previously delayed payments for depredation compensation; a lack of empathy or acknowledgement by a government official after the loss of a family member to a lion; previous exclusion from a decision-making process. Such past events may colour current events, adding resentment, distrust, fear and anger to the current situation, making it more difficult to resolve. Even if only one person in the dispute perceives that there were previous disputes that were not settled satisfactorily, underlying conflict exists and will distort the dynamics around the current incident. Parties to the conflict do not need to perceive the past similarly for an underlying conflict to exist.

The third level of conflict is identity conflict. Identity conflict occurs when assumptions, judgements or prejudices relating to a person's affiliation with a specific group impacts that person or group's receptivity or ability to address the conflict. Adversaries ascribe an identity to themselves and their opponents, each side thinking the fight is between 'us' and 'them' (Kriesberg 2003). Such conflicts transpire when a person or a group perceives that his or her sense of self or group is threatened, or deprived of legitimacy, respect or recognition. Identity-based conflicts can also originate from real or perceived power imbalances, disparate cultural or spiritual beliefs, repeated violations of basic dignity or a track record of resistance to collaboration.

Identity is so fundamental to a person or group's perceptions of themselves and the outside world that any real or perceived threat to identity can lead to intractable conflict (Burton 1990). Identity needs, unlike interests, are non-negotiable. Thus, situations that include identity conflict are particularly challenging to reconcile. Identity conflicts are significant contributors to 'wicked problems' (Rittel and Webber 1973: 155). Horst Rittel and Melvin Webber first described 'wicked problems' in the 1970s as 'problem[s] wrought with incomplete and contradictory information and changing requirements thus making it difficult to solve' (1973: 155). Wicked problems are broadly defined as environmental issues having a high level of scientific uncertainty and a diversity of interested party perspectives (Kreuter et al. 2004; Balint et al. 2011).

Values, beliefs, racial, gender, religious, ethnic and other inter-group conflict may enter the conservation context, in addition to differing views. Thus, identity conflict can arise in conservation in a multitude of dynamics. Experience suggests that there are few contexts where identity does not enter into the conservation conflict.

Unlike disputes, which tend to be concrete and physical, easily defined and articulated, identity-level conflicts tend to be intangible, ambiguous and unquantifiable. They are often either unclearly expressed or unappreciated in their links to the conservation context. Further, identity conflicts are often voiced as material disputes. Articulating an identity conflict as a dispute gives concreteness, clarity and focus to the group's concern (Rothman 1997). It is frequently more publicly acceptable to discuss material losses rather than social or psychological injuries involving status or dignity. The conservationist's interest in and focus on the wildlife tends to redirect conversations in a way that can appear to minimize a person's sense of psychological, social or cultural well-being.

Further confounding matters, a conflict that starts as a simple dispute may, over time, develop into an identity conflict, as the groups involved entrench themselves in the dispute level issue, such that the dispute becomes an essential part of their identity (Lederach 1997). The growing controversy over monk seals in Hawaii is an excellent example. Originally a pest to native fishermen, monk seals have evolved into a modern symbol of oppression and injustice committed by outsiders, particularly the US federal government. A recent surge in violent killings of monk seals has less to do with competition over fish, and more to do with an attempt to express outrage over past and current perceived injustices (Mooallem 2013).

Conflicts over wildlife can often be proxies for social conflicts involving a group's need for respect, recognition, autonomy or voice. Some of these may be in direct response to the presence of conservation authorities within the community, and some may have their roots elsewhere in society. While these unmet social and psychological needs may appear to be peripheral or unconnected to conservation, if these aspects of the conflict are not recognized, they will immediately or eventually impede conservation efforts.

By better understanding these deeper conflicts, conservationists are given a window into the reasoning behind seemingly irrational or counterproductive behaviour, for example when a community intentionally destroys, or fails to adopt, a conservation measure that would minimize wildlife damage to crops or livestock. Such an obstructionist display may be an expression of a deeper conflict and resentment over the unmet non-negotiable needs experienced within the local community. From a conservationist's point of view, the seemingly illogical behaviour is frustrating and disheartening. However, the CCT approach contends, counter-intuitively, that these incidents present conservationists with an opportunity to understand and harness the underlying social conflicts and group-recognition needs of the people in question. In doing so, the conservationist might change tactics and instead of immediately rebuilding the fence, focus more on rebuilding relationships or establishing reconciliation processes within the community. This would help to re-establish respect and trust and ensure that communities have appropriate ownership over and receptivity to decisions that impact them.

As the cases in this chapter illustrate, there is considerable potential for positive change when CCT principles and processes are engaged within a conservation conflict context. There is growing individual and organizational application of CCT in conservation by those who have participated in the HWCC's capacity-building workshops over the last seven years. Consistent anecdotal evidence indicates that CCT is more effective than the other stakeholder engagement and conflict approaches previously used in these projects and regions. Still, there is a need to systematically evaluate the multiple approaches being utilized in conservation to address conflict and stakeholder engagement to determine if CCT is indeed the more effective approach.

Implications of Underlying and Identity Conflict for Conservation

When underlying and identity conflict is present, the level of conflict is intense and complex and the reactions may appear 'irrational'. Conservation practitioners wonder why scientific data, economic reimbursement to offset damage by wildlife and community education programmes are met with resistance or even lead to an escalation of the conflict. Human-wildlife conflict researchers and practitioners often differentiate measured crop or livestock loss with the existence of seemingly exaggerated complaints by a farmer in assessing crops or livestock lost, labelling them as 'perceived' conflict (Naughton-Treves 1997). Their focus instead is on the 'real' conflict – that is, the wildlife crop or livestock damage that is tangible and measurable – which is typically far smaller by objective, material measure (Naughton-Treves 1997; Siex and Struhsaker 1999; Gillingham and Lee 2003; Madden 2004). Yet, if we were to broaden our definition of human-wildlife conflict to encompass all three levels of conflict, we would understand 'real' (or measured) conflict to be the dispute, and perceived conflict to be an expression of the intensity of underlying and identity conflict.

The processes used to address each level of conflict most effectively are different, with the deeper-rooted conflicts requiring a greater allocation of time, effort and resources aimed at the reconciliation and decision-making process. Conflicts, which arise over a violation of rights between two individuals or groups, can be settled relatively easily if the conflict is only at the dispute level. If a lion ate a cow and there is no other history at play, simply paying to mitigate the cost of losing the cow or erecting a fence to prevent future losses could settle the issue. Governments may establish laws to protect an endangered species with the courts deciding punishment for those who violate the law. Conservation groups may file a lawsuit to re-establish such designations if legal protections are removed. Yet much conflict in conservation extends beyond the realm of rights-based disputes; in fact compensation, fences, laws and lawsuits may actually inflame the underlying and identity conflicts, and subsequently provoke strong reactions, if the deeper conflicts are not being satisfactorily addressed (Ginges et al. 2007). This awareness lies at the centre of the CCT approach. If underlying and deep-rooted conflicts are present and not sufficiently addressed, any settlement is only temporary as those involved will use (or create) any opportunity to rectify previous injustices.

When underlying conflict is present, the negotiable needs and shared interests need to be identified, while trust and constructive communication needs to be re-established, before the conflict can be resolved. Many mainstream conflict-resolution, negotiation and participatory planning processes that have been adapted for conservation focus their analyses and processes, at least in part, and at best, at this level of conflict (Hannah et al. 1998; Llambi et al. 2005; Hammill et al. 2009; Treves, Wallace and White 2009; see also the Center for Natural Resources and Environmental Policy at University of Montana, www.cnrep.org).

Yet, most conservation struggles extend further to the third level of identity conflict. These deep-rooted conflicts involve non-negotiable needs. Any effort to engage opponents in a process that fails to reconcile these identity conflicts will sooner or later lead to failure. In fact, research suggests that attempting to address deep-rooted conflict through traditional negotiation means may actually intensify the conflict (Ginges et al. 2007). When identity conflict is present, a more involved reconciliation process is warranted. But the payoff when relationships are reconciled is that the conflict is transformed and the relevant parties often demonstrate a greater capacity for creativity and collaboration (Madden and McQuinn 2014).

The Three Facets of the Conflict Intervention Triangle Model

To provide a basic framework for incorporating the different levels of conflict in any conservation intervention, the following Conflict Intervention Triangle model advances the CCT framework by providing insights into what should be included.

As explained above, conflict can be multifaceted and daunting, particularly in conservation settings. HWCC's adaptation of the Conflict Intervention Triangle provides a framework for relating the three sources of conflict and corresponding points of intervention: process, relationships and substance (Moore 1986; Walker and Daniels 1997; Madden and McQuinn 2014). All elements must be integrated if conservation conflicts are to be successfully undertaken (see figure 8.2).

Of the three sets of factors aligned along the sides of the triangle in figure 8.2, 'substance' is the most straightforward and most proficiently addressed in the

Three Dimensions of Conflict

Process (how) • Substance (what)

Relationships (who)

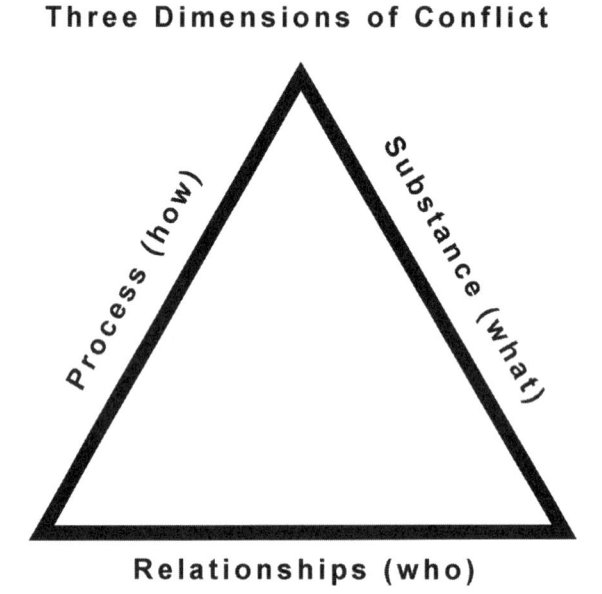

Figure 8.2. Conflict Intervention Triangle Model. *Source*: Madden and McQuinn (2014).

conservation field. These factors predictably parallel the issues contended for at the dispute level of the Levels of Conflict model. Typically, practitioners focus on the substantial, tangible concerns in conflict, often overlooking the full need of the process and relationship causes of conflict that we focus on in this section.

Limitations or failures in process are a considerable source of conflict in conservation. Such conflicts are largely focused on who makes the decision and how those decisions are made. A good process can improve chances of addressing social and psychological needs for recognition, meaningful participation, freedom of choice, identity and respect. If not addressed, conflict ensues or continues. Thus, stakeholders may approve of the merits of a specific solution, but if they do not feel they have contributed to the decision or that their needs and concerns were not fully recognized, they will share no sense of ownership over the decision and may ultimately reject or even sabotage any decision reached. It can be bewildering when stakeholders initially express agreement with, but then eventually reject, a solution that addresses their concerns. Such instances underscore the value of process in resolving conservation conflicts. Interestingly, process is also identified as of greater consequence in business decisions than is analysis of options (Lovallo and Sibony 2010: 6).

Dan Lovallo and Olivier Sibony (2010) further determined that decision-makers erroneously assume that the capacity to design and lead a decision-making process is implicitly understood and managed effectively. In fact, the skills and strategies for successfully designing and leading collaborative decision-making need to be learned and practiced, no less than what is expected within the natural sciences, where scientists must be trained in the correct methods for conducting research. We contend that conservation's long-term involvement to achieve and maintain social conditions favourable to human-wildlife coexistence makes conflict transformation a more appropriate model for addressing the ebb and flow of conflict in conservation.

The conflict transformation model has several commonalities with conflict resolution, an approach that is the basis of most conservation participatory stakeholder processes, but it differs in several key respects. Both approaches seek to remedy the immediate problem, but where conflict resolution is content-centred, conflict transformation is more relationship-centred. Conflict resolution may touch on underlying conflict in its efforts to seek resolution, but conflict transformation addresses all three levels of conflict with particular attention to identity-based conflicts, given their near universal presence and capacity for ensuring intractability. Conflict resolution is more geared toward short-term engagement and agreement on a solution. Conflict transformation seeks to put in place processes that address current problems through a mid- to long-term process that promotes change in the underlying structures that drive the creation of recurrent conflict. In other words, while both approaches seek to end something not desired, conflict transformation also seeks to build something constructive in its place. In essence, conflict transformation uses efforts to address the current problem as an entry point to both reconcile relationships and foster long-term positive change in the social system (Lederach 2003; Lederach, Neufeldt and Culbertson 2007). An implication of the CCT framework is a re-examination by the conservation field of what effective stakeholder engagement actually means.

Unfortunately, most current conservation stakeholder engagement processes are often singularly focused on 'the big meeting'; driven toward quick decisions, often at the expense of relationships; and structured to highlight power imbalances and disparities in decision-making authority. As a result, solutions that may have been agreed upon tend to collapse from a lack of sustained support, and short-term gains are eventually overshadowed by long-term losses. Models, such as that proposed by Steve Redpath et al. (2013) offer a step-wise process of managing conflict that is largely linear and engages the stakeholders too late in the design process. The process may even prevent stakeholder receptivity from being built because of the limits and timing of engagement and the fact that engagement is determined with a simple yes/no query as to whether or not the stakeholder wishes to engage. Such an approach may increase the likelihood of failure because it misses out on important opportunities to build respect, trust and empowerment, emphasizes one process throughout, and fails to allow adequate time and opportunity to identify common ground and create a collective definition of the problem. Consequently, such approaches fail to create conditions for stakeholder receptivity, creativity and flexibility not only before or during the 'big meeting', but also throughout the much longer phase of implementation and continued adaptation (Lederach 2003).

Given the pattern of underwhelming outcomes from past stakeholder engagement efforts, it is understandable that conservationists may be resistant to relinquishing decision-making control to stakeholders who seem less committed, or even antagonistic toward conservation objectives. The fear expressed by conservationists is simple: if we involve other stakeholders in the decision-making around wildlife, we will have to compromise in conserving species and spaces. However, CCT processes allow for an expansion of the decision-making space, creating opportunities to address the social and psychological needs of stakeholders that frequently inhibit their willingness to be flexible and receptive to others' needs. As a result, CCT processes often result in conservationists having to compromise less about the issues they hold most dear. Indeed, successful decision-making processes have other positive outcomes besides sustainable solutions. These include the establishment of healthy relationships that better prepare conservation partners to anticipate and address future conflicts and efficiently adapt solutions to changing social and environmental conditions. These changes in process and relationships, as described below, also have the benefit of allowing individual stakeholders to make better, more informed decisions within their organization or group (Madden and McQuinn 2014).

The third source of conflict and corresponding point of intervention is relationships, both interpersonal and intergroup. Disrespect and distrust between two or more people or groups can impede efforts to understand the truth about a conflict or undermine efforts to establish sustainable coexistence between people and wildlife. In our experience, this source of conflict is too often ignored, avoided or treated too lightly by conservation and government authorities who label other groups as 'partners in conservation' when that relationship is still wrought with distrust and efforts to undermine the goals of coexistence. Experience suggests that stakeholders will undervalue or even sabotage conservation solutions if they do not also sustain relationships characterized by respect and trust. In fact, for coexistence to be authentically achieved, it is critical that antagonists in a conflict are able to envision the essential interdependence each side has with the other (Lederach 2005). The following cases illustrate the value of investing in process and relationships as a means to transform conservation conflicts.

Case Studies in Conservation Conflict Transformation
Case Study 1: Poaching in Mozambique

In Mozambique, a conservation project leader had started an anti-poaching scouting programme in an area rampant with poaching, in large part due to local Chinese influence and demand for ivory. On the day the first graduation class of scouts was presented to the local villages, one village chief spoke eloquently about the importance of conservation and an end to poaching. The next day, the scouts went out on their first official patrol and found snares. They followed the trail right back to the same local chief who the day before spoke of the importance of conservation and anti-poaching. When threatened with arrest, the chief in turn threatened the lives of the scouts and their families through shamanistic magic (which in this culture is feared more than a direct physical threat). The conservation leader was in a difficult bind. If she allowed the chief to get off without punishment, her project's credibility would be undermined/damaged. If she arrested the chief, lives could be at stake and the chief would turn against both conservation and her project.

Relying on the principles of conflict transformation, she facilitated the dialogue with the community. Instead of focusing on who was to blame or offering suggestions for solutions, she facilitated the dialogue in a positively centred way and empowered the initially angry, emotionally charged community members to determine for themselves the best course of future action. There was a long, awkward silence. Finally, another chief proposed an amnesty, suggesting the communities needed some time to get used to this idea of no poaching. He suggested they have a period of two weeks where everyone would turn in their snares and then, after that, anti-poaching scouting would begin. A few days after the meeting, the offending chief turned in four snares and during the two-week amnesty, others turned in ten more. This effort by the community to turn in snares continued beyond the proposed amnesty and grew to a community-wide effort to seek out and arrest poachers. The conservation project and its staff maintained their project's credibility and the chief was able to save face. Moreover, the solution was truly the local people's solution and they genuinely owned it, something that would not likely have been achieved if the conservation leader had suggested the solution herself. By her own admission, she reported that this was a much better solution than any she had thought of on her own. Indeed, when the conservation authority withholds the temptation to impose solutions, opting instead to create the conditions for an empowered transformational process, the community makes decisions that they are more likely to genuinely and sustainably own (Beggs pers. comm. 2012).

Case Study 2: Creating a Conservation Conflict to Secure Native Identity and Retaliate Against the Government

In another case, stakeholders on a certain island came together to build their capacity to address local conservation conflicts (note: details of this case are omitted to maintain confidentiality around this highly sensitive conflict). The conflict involved an exotic, invasive ungulate species that was illegally introduced to the island – an act that could destroy both the cattle ranching industry and native flora and fauna, if eradication efforts were not jointly and immediately undertaken. However, given a long and tumultuous history on the island, the native people's distrust of mainland conservation organizations and government-supported invasive species efforts was an overriding obstacle to making progress. In fact, conservation advocates had failed to garner support from ranchers for eradication, despite the mutual recognition that a burgeoning ungulate population would likely put the ranchers permanently out of business. Over the course of four days, enough trust was built between individuals from the two groups that native stakeholders with ranching interests revealed that the reason the invasive ungulates were illegally and surreptitiously being brought to the island was to redirect conservation's human and financial resources away from eradication of feral pigs and sheep, which are culturally important for traditional hunting practices, and perhaps more so to serve as a 'poke-eye' to the national government's perceived arrogance. To date, immense resources were, indeed, being redirected and funnelled into eradicating the invasive ungulates, reducing the amount of resources available for other invasive species. What was revealed through this process was that maintaining the cultural security to hunt – a critical human need for these people – and overcoming the perception of government arrogance together outweighed even the likely economic destruction and loss of ranching as a livelihood. And yet, the trust developed between individuals from these two groups was enough to diminish the threat to the native people's identity so that a critical truth could be revealed.

Case Study 3: Reconciliation of Deep-Rooted Conflict Leads to Great Creativity and Collaboration

Another recent multi-stakeholder capacity-building and planning workshop brought together staff from a federal government agency in the United States, tasked with providing the option of lethal interventions in cases of human-wildlife conflict, and two conservation groups whose strategies emphasized non-lethal rather than lethal control of wolves. The wolf-conservation groups' orientation to the current conflict was to blame the federal agency for an overuse of lethal control of wolves among ranchers. By law, if there was a depredation incident, the rancher could call upon federal agency staff to lethally remove the wolf. Wolf-conservation advocates felt that this agency's over-reliance and excess provision of lethal alternatives was impeding their efforts to increase uptake of non-lethal methods by ranchers. Prejudicial assumptions about the 'other' were held on both sides. Over several days, individuals from the federal agency and the NGOs worked together through a capacity-building workshop so they could better analyse and address their conflict. Through this process, the participants humanized one another and began to build a positive relationship. They engaged in small, shared wins that built their confidence and capacity to work together more effectively. In less than a week, these frequent adversaries developed respect and trust with each other and built a shared capacity and motivation to transform their conflict.

Towards the end of the workshop, the non-lethal control advocates shared their conflict analyses with the whole group. The analyses included levels of conflict and a conflict map that visually depicted, among other things, the concern that the federal agency was enabling an overuse of lethal control for wolf depredation. Using that analysis, it was the staff of the federal agency that not only proposed creative ideas for improving rancher openness to non-lethal techniques, but also proposed ways that their agency could facilitate an increase in receptivity by ranchers to non-lethal alternatives. The conflict transformation process allowed both sides to reconcile their relationships and move beyond institutional and identity prejudices and narrow assumptions of the other, allowing for a more creative, transformative set of solutions to emerge. This case, while only a microcosm of the institutionalized, system-level conflict involving these parties, gives some indication of what might be possible if efforts were made on a larger scale.

Case Study 4: The Power of Listening, Respect and Acknowledgment to Build Trust and Create Coexistence
A woman, who runs a predator-conservation organization in northern Kenya, applied conflict transformation principles and skills to a specific one-on-one conflict, and achieved a positive outcome. In 2012, an old man from the local village came to her conservation camp carrying the bloody legs of the eight goats recently killed by a pack of endangered African wild dogs. One of the goats was a nanny goat that provided milk for a substantial part of the village. He was angry and upset. He demanded compensation while also asserting his intention to kill the pack of endangered predators. Rather than trying to 'solve' his problem by suggesting changes in livestock husbandry, warning him that killing a highly endangered species is illegal or informing him that compensation for losses were unfortunately not available, she recognized his need to be truly heard and acknowledged for his loss of the goats. By simply listening and recognizing his loss, she addressed a much deeper human need for social recognition and respect. She conveyed that this man was at least as important as the wildlife in that conversation and that his loss was her concern. After their conversation, the man left her camp. He was still saddened by his loss, certainly, but he was no longer angry and insistent on compensation or retaliation. A couple of months later, once he had re-established his herd of goats, he brought her a goat as a present. Why? Because despite his relatively impoverished position in life, his unstated need for respect, recognition and acknowledgment were far greater than his spoken need for economic remuneration and revenge. As this example illustrates, the basic building block of any transformative change is at the interpersonal level, however large the number of individuals or groupings involved in a conflict. Further, though this seems like a simple task, the skills are challenging to master and apply consistently in a high-tension scenario. They require targeted capacity building and continual reflective practice (Bhalla pers. comm. 2012).

Important Gaps in Current Conservation Capacity

Wildlife conservation professionals characteristically enter the field with a focus on the wildlife – not humans. Their principal interest is in conserving wildlife and engaging in research related to the wildlife or ecosystems they aspire to conserve. While educational opportunities in 'human dimensions' have significantly increased in recent years (Robertson and Butler 2001; Decker, Riley and Siemer 2012b), graduates in this sub-field often lack the specific skills required to analyse conflict successfully or to design and lead effective interventions to address it, let alone the practical skills of how to transform social conflict into productive opportunities for coexistence (Sample et al. 1999; Bonine, Reid and Dalzen 2003; Jacobson 2010). Indeed, most conservation professionals have not been provided with the training needed to work successfully with people and groups that may feel angry or marginalized because of past or present conservation decisions, not to mention the threats those institutions and decisions may signify to their identity and way of life (Bonine, Reid and Dalzen 2003; Madden 2004; Lederach, Neufeldt and Culbertson

2007; Madden 2008; Jacobson 2010). Even where training is geared toward conflict management and stakeholder engagement, often the focus is on efforts to determine shared interests that may be inaccessible or not able to be realized if identity conflict is at the root of the conflict. The need for public participation, community involvement, multi-stakeholder processes and transparent decision-making are often acknowledged, but training often gears the student toward linear, rational decision-making procedures that overly focus on 'the big meeting' as described in Redpath et al. (2013). As a result, they may neglect the steps needed to create conditions for social receptivity or to reconcile identity-based conflict. These approaches have utility in limited circumstances, but for most wildlife conservation contexts, they are insufficient (Lederach 2003).

The CCT approach advances efforts toward developing field-wide capacity to analyse the multifaceted drivers of deep-rooted social conflicts, and a foundation in transformational processes that can disentangle intractable conflicts rooted in an 'us' versus 'them' mind-set. Such processes give equal consideration to the dialogue and relationship building needed to foster dignity, respect and trust, as well as fostering more effective decision-making around concrete solutions. Certainly, some practitioners manage at times to do this, but typically work intuitively and over years of trial and error, as the field offers no framework for them to share their knowledge explicitly. Moreover, there is no accepted standard of practice for what it means to transform stakeholder conflict sustainably, and as such some efforts might be successful, while others may create more conflict than they resolve. This severely limits the transferability of their success and the extent to which good practices and principles can be incorporated into training, education or professional exchanges, or ideally passed from one generation of conservation practitioners to another.

Implications and Recommendations for the Conservation Field

As an essential first step, conservation projects need systematic inclusion of early and ongoing analysis of the social conflict dynamics to understand the human influence in creating and exacerbating conflict, avoid unintended consequences and foster social conditions that support constructive decision-making toward sustainable conservation (Madden 2004; Madden and McQuinn 2014). Second, conservation organizations and practitioners need a greater reliance on transformative decision-making processes that are designed to use the dispute-level issues as entry points to transform the underlying and identity conflict among stakeholders (Lederach 1997; Lederach 2003).

There is an understandable desire among conservationists to act rapidly to protect species that seem on the brink of extinction in the wild. And in many cases, short-term action is not only needed, but it can communicate a good-faith effort to those negatively impacted by wildlife. But such actions cannot stand alone. The CCT approach advocates 'going slow to go fast', recognizing that a more methodical, comprehensive approach to acknowledge the full spectrum of human needs will create sustainable support for coexistence in the long term (Fisher et al. 1991).

To that end, planning for how stakeholders are engaged is at least as important as the act of bringing the conflicting stakeholders together. We argue that this step is critical to the success of multi-stakeholder dialogues. Finally, it is the emphasis on relationship-building through transformative process design and mid- to long-term implementation and adaptation that we believe will aid conservation in achieving sustainable success (Lederach, Neufeldt and Culbertson 2007; Madden 2012).

The ambition of conflict transformation is to foster the conditions for sustainable human-wildlife coexistence, not simply create short-term solutions for, or superficial compliance with, conservation objectives (Lederach 2007; Madden and McQuinn 2014). CCT not only strives to change negative relationships between the conflicting parties but also to change the social and political structures and relationships impacting the achievability of conservation efforts in the longer term. CCT empowers stakeholders in conservation to become involved in dynamic constructive change processes to help build sustainable and equitable conditions for coexistence.

Implementation of these recommendations will require additional resources to build the capacity of conservation leaders and practitioners, as well as stakeholders, to understand and utilize the skills, principles and processes of CCT. But doing so will likely mean that conservation projects are more effective and efficient on the ground and more strategically positioned within the larger social system to maximize their efficacy. In addition, it is essential that conservation efforts adjust their focus to give equal emphasis to the substance, relationship and process aspects of conservation projects. The founding of the HWCC represented a critical and promising first step in the process of building and integrating CCT capacity, facilitating transformational interventions and establishing a standard of practice in how the conservation field addresses human-wildlife and other conservation conflicts.

The cases described above suggest how our field might evolve to increase the receptivity of people for long-term coexistence more strategically and efficiently. And while additional research to evaluate the efficacy of CCT in conservation is needed, preliminary research and experience suggests it can make a difference (Rothman 1997; Madden 2004; Ellis et al. 2005; Hendrick 2009; Madden 2012). Through this approach, the field may begin to re-align the underlying structures of conflict within complex social systems to support a change in attitudes, relationships and behaviours, leading to positive social change and sustainable coexistence. The future of conservation for many wildlife species relies not just on innovative solutions, but also on an increased tolerance and social carrying capacity that cannot be achieved by laws, science, money or fences alone.

Francine Madden is the co-founder and executive director of the Human-Wildlife Conflict Collaboration – a non-profit organization integrating 'conservation conflict transformation' strategies in wildlife conservation efforts. Francine has successfully facilitated conflict intervention, planning and capacity-building processes in some of the world's most fragile hotspots. Francine has helped people and projects significantly curtail wildlife poaching and trafficking, reconcile fractured relationships and dramatically improve overall social receptivity toward decision-making for wildlife conservation on every continent where humans and wildlife coexist. Francine Madden has two masters' degrees from Indiana University and is the author of numerous publications and presentations.

Brian McQuinn, a Harry Guggenheim Foundation Dissertation Fellow, is a Postdoctoral Researcher at the University of Oxford's Institute of Cognitive and Evolutionary Anthropology. He is also a Research Associate at the Centre on Conflict, Development and Peacebuilding and the Danish Institute for International Studies. Brian has numerous publications and his research has appeared in a range of media. Brian worked for fourteen years in more than a dozen conflict-affected countries as a dialogue specialist with the United Nations and other international organizations.

Note

1. For further information on the HWCC and its founding organizations, see http://www.humanwildlifeconflict.org/.

References

Anderson, M.B., and L. Olson, with assistance from K. Doughty. 2003. *Confronting War: Critical Lessons for Peace Practitioners*. Cambridge, MA: The Collaborative for Development Action, Inc.

Balint, P., et al. 2011. *Wicked Environmental Problems: Managing Uncertainty and Conflict*. Washington, DC: Island Press.

Bonine, K., J. Reid and R. Dalzen. 2003. 'Training and Education for Tropical Conservation', *Conservation Biology* 17(5): 1209–1218.

Breitenmoser, U.A., et al. 2005. 'Non-Lethal Techniques for Reducing Depredation', in R. Woodroffe, S. Thirgood and A. Rabinowitz (eds), *People and Wildlife, Conflict or Coexistence?* Cambridge: Cambridge University Press.

Burton, J.W. 1984. *Global Conflict*. Brighton: Wheatsheaf.

———, ed. 1990. *Conflict: Human Needs Theory*. New York: St. Martins Press.

Canadian Institute for Conflict Resolution. 2000. *Becoming a Third-Party Neutral: Resource Guide*. Ottawa: Ridgewood Foundation for Community-Based Conflict Resolution (Int'l).

Coate, R.A., and J.A. Rosati. 1988. *The Power of Human Needs in World Society*. Boulder, CO: Lynne Rienner Publishers.

Decker, D.J., S.J. Riley and W.F. Siemer. 2012a. 'Introduction to Human Dimensions of Wildlife Management', in D.J. Decker, S.J Riley and W.F. Siemer (eds), *Human Dimensions of Wildlife Management*, 2nd ed. Baltimore, MD: Johns Hopkins University Press, pp. 3–14.

———. 2012b. 'Adaptive Value of Human Dimensions for Wildlife Management', in D.J. Decker, S.J. Riley and W.F. Siemer (eds), *Human Dimensions of Wildlife Management*, 2nd ed. Baltimore, MD: Johns Hopkins University Press, pp. 248–256.

Dickman, A.J. 2010. 'Complexities of Conflict: The Importance of Considering Social Factors for Effectively Resolving Human-Wildlife Conflict', *Animal Conservation* 13(5): 458–466.

Ellis, C., et al. 2005. 'Approaching the Table: Transforming Conservation-Community Conflict into Opportunity', *Beyond the Arch: Community and Conservation in Greater Yellowstone and East Africa*. 7th Biennial Scientific Conference on the Greater Yellowstone Ecosystem, Mammoth Hot Springs, Yellowstone National Park, October 6–8. Proceedings edited by A.W. Biel.

Fisher, R., W. Ury and B. Patton. 1991. *Getting to Yes: Negotiating Agreement Without Giving In*, 2nd ed. New York: Penguin Books.

Frahm, J., and K. Brown. 2007. 'First Steps: Linking Change Communication to Change Receptivity', *Journal of Organizational Change* 20: 370–387.

Gillingham, S., and P.C. Lee. 2003. 'People and Protected Areas: A Study of Local Perceptions of Wildlife Crop-Damage Conflict in an Area Bordering the Selous Game Reserve, Tanzania', *Oryx* 37(3): 316–325.

Ginges, J., et al. 2007. 'Sacred Bounds on Rational Resolution of Violent Political Conflict', *Proceedings of the National Academy of Sciences of the United States of America* 104(18): 7357–7360.

Hammill, A., et al. 2009. *Conflict-Sensitive Conservation: Practitioners' Manual*. Winnipeg: International Institute for Sustainable Development.

Hannah, L., et al. 1998. 'Participatory Planning, Scientific Priorities, and Landscape Conservation in Madagascar', *Environmental Conservation* 25(1): 30–36.

Hendrick, D. 2009. *Complexity Theory and Conflict Transformation: An Exploration of Potential Implications*. Working Paper 17. Centre for Conflict Resolution, Department of Peace Studies, University of Bradford.

Human-Wildlife Conflict Collaboration (HWCC), 2008. 'Benefitting Conservation Through Conflict Transformation.' White paper, Washington, DC.

IUCN. 2003. 'World Parks Congress Recommendation: Preventing and Mitigating Human-Wildlife Conflicts'. Reprinted in *Human Dimensions of Wildlife* 9: 259–260.

Jacobson, S.K. 2010. 'Effective Primate Conservation Education: Gaps and Opportunities', *American Journal of Primatology* 72(5): 414–419.

King, L.E., I. Douglas-Hamilton and F. Vollrath. 2011. 'Beehive Fences as Effective Deterrents for Crop-Raiding Elephants: Field Trials in Northern Kenya', *African Journal of Ecology* 49: 431–439.

Kreuter, M.W., et al. 2004. 'Understanding Wicked Problems: A Key to Advancing Environmental Health Promotion', *Health Education & Behavior* 31(4): 441–454.

Kriesberg, L. 2003. 'Identity Issues', in G. Burgess and H. Burgess (eds), *Beyond Intractability*. Conflict Information Consortium, University of Colorado, Boulder. Retrieved 15 January 2008 from http://www.beyondintractability.org/bi-essay/identity-issues.

Lederach, J.P. 1997. *Building Peace: Sustainable Reconciliation in Divided Societies*. Washington, DC: United States Institute for Peace.

———. 2003. *Little Book of Conflict Transformation*. Intercourse, PA: Good Books.

———. 2005. *Moral Imagination: The Art and Soul of Building Peace*. Oxford: Oxford University Press.

Lederach, J.P., R. Neufeldt and H. Culbertson. 2007. *Reflective Peacebuilding: A Planning, Monitoring and Learning Toolkit*. The Joan B. Kroc Institute for International Peace Studies, University of Notre Dame and Catholic Relief Services (CRS).

Llambí, L.D., et al. 2005. 'Participatory Planning for Biodiversity Conservation in the High Tropical Andes: Are Farmers Interested?' *International Mountain Society. Mountain Research and Development* 25(3): 200–205.

Lovallo, D., and O. Sibony. 2010. 'The Case for Behavioral Strategy', *McKinsey Quarterly*. Boston: McKinsey. Retrieved 13 April 2010 from http://www.mckinsey.com/business-functions/strategy-and-corporate-finance/our-insights/the-case-for-behavioral-strategy.

Madden, F. 2004. 'Creating Coexistence between Humans and Wildlife: Global Perspectives on Local efforts to Address Human-Wildlife Conflict', *Human Dimensions of Wildlife* 9: 247–257.

———. 2006. 'Gorillas in the Garden: Human-Wildlife Conflict in Bwindi Impenetrable National Park', *IUCN Policy Matters 14*, March.

———. 2008. 'Human-Wildlife Conflict: A Case for Global Collaboration', Pathways to Success: Human Dimensions of Fish and Wildlife Management Conference, Colorado, 15–19 September.

———. 2012. 'Conflict Transformation: Untangling Conservation Conflict in Contemporary Society', Pathways to Success: Human Dimensions of Fish and Wildlife Management Conference, Colorado, 23–28 September.

Madden, F., and B. McQuinn. 2014 'Conservation's Blind Spot: The Case for Conflict Transformation in Wildlife Conservation', *Biological Conservation* 178: 97–106.

Manfredo, M.J., et al. 1998. 'Public Acceptance of Mountain Lion Management: A Case Study of Denver, Colorado, and Nearby Foothills Areas', *Wildlife Society Bulletin* 26(4): 964–970.

———. 1999. 'Public Acceptance of Wildlife Trapping in Colorado', *Wildlife Society Bulletin* 27: 499–508.

Manfredo, M.J., and A.A. Dayer. 2004. 'Concepts for Exploring the Social Aspects of Human–Wildlife Conflict in a Global Context', *Human Dimensions of Wildlife: An International Journal* 9(4): 1–20.

Manfredo, M.J., T.L. Teel and A.D. Bright. 2003. 'Why are Public Values toward Wildlife Changing?' *Human Dimensions of Wildlife* 8: 287–306.

Marker, S. 2003. 'Unmet Human Needs', in G. Burgess and H. Burgess (eds), *Beyond Intractability*. Conflict Information Consortium, University of Colorado, Boulder. Retrieved 15 January 2008 from http://www.beyondintractability.org/essay/human-needs.

Maslow, A. 1954. *Motivation and Personality*. Reading: Addison-Wesley Publishing Company.

Max-Neef, M.A., A. Elizalde and M. Hopenhayn. 1989. 'Development and Human Needs', in *Human Scale Development: Conception, Application and Further Reflections*. New York: Apex, p. 18.

McShane, T.O., et al. 2011. 'Hard Choices: Making Trade-offs between Biodiversity Conservation and Human Well-Being', *Biological Conservation* 144(3): 966–972.

Mooallem, J. 2013. 'Who Would Kill a Monk Seal?' *New York Times*, 8 May. Retrieved 10 May 2013 from http://nyti.ms/1Gvm9dw.

Moore, C.W. 1986. *The Mediation Process: Practical Strategies for Resolving Conflict*. Hoboken, NJ: Wiley.

Naughton-Treves, L. 1997. 'Farming the Forest Edge: Vulnerable Places and People Around Kibale National Park, Uganda', *American Geographical Society Geographical Review* 87(1): 27–46.

Nie, M.A. 2003. *Beyond Wolves: The Politics of Wolf Recovery and Management*. Minneapolis: University of Minnesota Press.

Nyhus, P.J., et al. 2005. 'Bearing the Costs of Human-Wildlife Conflict: the Challenges of Compensation Schemes', in R. Woodroffe, S. Thirgood and A. Rabinowitz (eds), *People and Wildlife, Conflict or Coexistence?* Cambridge: Cambridge University Press, pp. 107–121.

Packer, C., et al. 2013. 'Conserving Large Carnivores: Dollars and Fences', *Ecology Letters* 16(5): 635–41.

Pidot, L. 2008. *Looking Beyond the Agency: The Influence of Stakeholder Engagement on the Perceived Success of the Maine, New Hampshire, and Vermont State Wildlife Action Plans.* School of Natural Resources and Environment, University of Michigan.

Redpath, S.M., et al. 2013. 'Understanding and Managing Conservation Conflicts', *Trends in Ecology & Evolution* 28(2): 100–109.

Reed, M.S. 2008. 'Stakeholder Participation for Environmental Management: A Literature Review', *Biological Conservation* 141: 2417–2431.

Richardson, D.M., et al. 2009. 'Multidimensional Evaluation of Managed Relocation', *Proceedings of the National Academy of Sciences of the United States of America* 106(24): 9721–9724.

Rittel, H.W.J., and M.M. Webber. 1973. 'Dilemmas in a General Theory of Planning', *Policy Sciences* 4: 155–169. Retrieved 25 April 2013 from http://www.uctc.net/mwebber/Rittel+Webber+Dilemmas+General_Theory_of_Planning.pdf.

Robertson, R.A., and M.J. Butler. 2001. 'Teaching Human Dimensions of Fish and Wildlife Management in U.S. Universities', *Human Dimensions of Wildlife* 6: 67–76.

Rothman, J. 1997. *Resolving Identity-Based Conflict: In Nations, Organizations, and Communities.* San Francisco, CA: Jossey-Bass Publishers.

Sample, V.A., et al. 1999. 'Forestry Education: Adapting to the Changing Demands on Professionals', *Journal of Forestry* 97(9): 4–10.

Siex, K.S., and T.T. Struhsaker. 1999. 'Colobus Monkeys and Coconuts: A Study of Perceived Human-Wildlife Conflicts', *Journal of Applied Ecology* 36(6): 1009–1020.

Sitati, N.W., and M.J. Walpole. 2006. 'Assessing Farm-Based Measures for Mitigating Human-Elephant Conflict in Transmara District, Kenya', *Oryx* 40: 279–286.

Smith, K.L., and C.B. Torppa. 2010. 'Creating the Capacity for Organizational Change: Personnel Participation and Receptivity to Change', *Journal of Extension* [On-line], 48(4) Article 4FEA1. Available at: http://www.joe.org/joe/2010august/a1.php.

Treves, A., R.B. Wallace and S. White. 2009. 'Participatory Planning of Interventions to Mitigate Human-Wildlife Conflict', *Conservation Biology* 23(6): 1577–1587.

Walker, G., and S. Daniels. 1997. 'Foundations of Natural Resource Conflict: Conflict Theory and Public Policy', in B. Solberg and S. Miina (eds), *Conflict Management and Public Participation in Land Management. EFI Proceedings 14, European Forest Institute*, pp. 13–36.

Woodroffe, R., S. Thirgood and A. Rabinowitz (eds). 2005. *People and Wildlife, Conflict or Coexistence?* Cambridge: Cambridge University Press.

9
Engaging Farmers and Understanding Their Behaviour to Develop Effective Deterrents to Crop Damage by Wildlife

Graham E. Wallace and Catherine M. Hill

The interests of humans and wildlife frequently conflict wherever they coexist (see Conover 2002; Madden 2004; Thirgood, Woodroffe and Rabinowitz 2005; Young et al. 2005). Consequently, understanding and addressing these 'conflicts' is a key management issue for wildlife conservation (Messmer 2000; Paterson and Wallis 2005; Woodroffe, Thirgood and Rabinowitz 2005; Dickman 2010). Perceptions of interactions between humans and wildlife typically reflect the extent to which individuals live with wildlife (Newmark et al. 1993; Rao and McGowan 2002; Altrichter 2006) and reconciling perspectives usually involves meshing disparate goals (Adams 1998; Abbot et al. 2001; Hutton and Leader-Williams 2003; Baker et al. 2013). Land use and access to resources are often central to human-wildlife interactions (see Blomley 2000; Kagiri 2002; Mukherjee and Borad 2004), particularly in many rural areas of Africa where increasing human populations depend on limited land and decreasing forest resources for livelihoods (Kepe, Cousins and Turner 2001; Bush et al. 2004; Plumptre et al. 2004; Toutain, De Visscher and Dulieu 2004). Resource competition between humans and wildlife can be prevalent where human communities relying on agriculture for food security and income adjoin forest reserves that provide essential habitat for wildlife (Hill 1997; Naughton-Treves 1997; Tungittiplakorn and Dearden 2002; Knickerbocker and Waithaka 2005). Crops near forest may be a predictable and accessible source of nutrition for wildlife, and therefore many species forage on them.[1] This can result in extensive ongoing damage to crops, with adverse impacts on farmer livelihood (Nyhus, Tilson and Sumianto 2000; Hill 2004; Chiyo et al. 2005; Rode et al. 2006).

Animals have probably entered farmers' fields to consume crops for centuries, but interest in assessing the impacts of this behaviour is relatively recent (Hill, Osborn and Plumptre 2002). Many accounts have approached crop damage by wildlife from a conservation perspective, focusing particularly on its implications for wildlife (Horrocks and Baulu 1988; Plumptre and Reynolds 1994; Hill 2002b; Choudhury 2004; Sitati and Walpole 2006). However, because living closely with wildlife often affects humans adversely it is now widely acknowledged that human perspectives must be taken into account to not only mitigate crop losses and associated livelihood impacts but also promote conservation (Soto, Munthali and Breen 2001; Conover 2002; Hill 2002a; Sitati and Walpole 2006). This includes addressing the direct and opportunity costs of wildlife crop damage for farmers and their communities (see Naughton-Treves 1997; CARE et al. 2003; Hill 2005; Priston and McLennan 2013). While these communities are often sympathetic to conservation goals, the conflicts and costs generated by crop damage can reduce tolerance of wildlife and undermine management plans (Infield 1988; Strum 1994; Happold 1995; Lee and Priston 2005; Baker et al. 2013). Hence it is critical to work in partnership with these stakeholders when developing and implementing strategies to mitigate crop losses and associated conflicts (Osborn and Hill 2005; Sitati, Walpole and Leader-Williams 2005; Graham and Ochieng 2008; Redpath et al. 2013). Levels of crop damage, costs to replace lost food and farmers' investments in time or resources to protect their crops are measures of the impact of crop damage on livelihoods (Colfer, Wadley and Venkateswarlu 1999; Hill, Osborn and Plumptre 2002; Bush et al. 2004; MacKenzie and Ahabyona 2012). Although the threshold levels of loss that undermine farmer livelihood security probably vary across time and sites, relatively minor crop loss can adversely affect subsistence households in developing countries because farmers often lack savings, crop surpluses or alternative sources of income to absorb and offset losses (Barbier 1987; Van Huis and Meerman 1997; Bush et al. 2004; Hill 2004).

It should be noted that while much of the existing literature on crop damage by wildlife uses the term 'crop-raiding', we avoid using it here. As discussed elsewhere, terms such as 'raid', 'crop-raiding' and 'crop-raider' imply an unauthorised, malicious or harmful attack on farmers' property, with negative connotations that do little to mitigate conflict (see Webber 2006; Peterson et al. 2010; Hill 2015).

To manage crop damage by wildlife it is essential to have an accurate and comprehensive account of crop-foraging activity and impacts, which includes understanding farmers' perceptions and behaviour, the behavioural ecology of the species concerned and patterns and parameters of crop-foraging events. Mitigation strategies should not only deliver reduced crop loss for farmers but also lessen the associated social costs of these events (Gillingham and Lee 2003; Hill 2005; Webber 2006) and increase tolerance of wildlife (Naughton-Treves 1998). It is also possible that crop damage is only one aspect or focus of wider farmer frustrations (Hill 2004; Madden

2004). It is therefore necessary to balance farmers' perceptions with what is observed to occur on farms, because disparity can compromise mitigation efforts; there will be little value in addressing A if the real cause or issue is B (see Hill, Osborn and Plumptre 2002; Osborn and Hill 2005).

Actively involving all stakeholders throughout each stage of the process of 'conflict' mitigation (1) improves understanding of perceptions and any differing perspectives, (2) ensures that aims, actions and outcomes are likely to be relevant across all stakeholders and aligned with their interests or concerns, (3) ensures interventions are locally appropriate and consistent with local norms and customs, (4) increases ownership of core issues and the need to find solutions and (5) maximizes the probability that strategies will be effective, and hence used or adapted, over the long term (Strum 1986; Kapila and Lyon 1994; Gillingham and Lee 1999; Hackel 1999; Miller and McGee 2001; Hill, Osborn and Plumptre 2002; Armitage 2003; Osborn and Parker 2003; Riley et al. 2003; Chase, Decker and Lauber 2004; Osborn and Hill 2005; Sitati and Walpole 2006; Webber, Hill and Reynolds 2007). Although many techniques traditionally employed by farmers to protect their crops from wildlife are not legally sanctioned, it is unreasonable and impractical to expect farmers to stop using methods they perceive to be effective without providing access to alternatives aligned with conservation goals (Hill, Osborn and Plumptre 2002).

Central to reducing crop losses due to wildlife foraging activities is ensuring farmers have access to a range of effective and affordable tools to protect their crops (Hill, Osborn and Plumptre 2002; Osborn and Hill 2005; Hill and Wallace 2012). To develop these tools it is necessary to measure and understand animal foraging activity within an area, including a thorough record of the frequency of crop-feeding events, which species feed on crops and which crops are damaged (Hygnstrom, Timm and Larson 1994; Conover 2002). It is also essential to understand when and where crop-foraging events occur, under what conditions they occur, the amount of crop loss that occurs and which parameters of crop-foraging events determine crop loss (Wallace and Hill 2012). Behavioural information about crop-foraging species can indicate how they respond or adapt to human activity and crop-protection efforts (Barnes 1996; Boydston et al. 2003; Beale and Monaghan 2004; Fuentes, Southern and Suaryana 2005). This information can then be used to design deterrents to crop foraging that address specific attributes of foraging dynamics as well as animal species characteristics (Conover 2002; Wallace 2010).

Deterrents are usually grouped according to how they operate or are used (for example, active, passive, physical, tactical, vigilant, noise, sensory, barrier, alarm, repellent, lethal); in practice, most can be assigned to more than one group. Some methods, such as burning pepper dung to deter elephants (Osborn and Parker 2002a), or hanging monofilament lines and flags to deter ducks (Lane and Higuchi 1998), are relatively species-specific, while methods such as chasing, guarding and using traps or wire fences may deter a broad range of species. Lethal removal of animals can occur through hunting or use of traps, snares, weapons, baits or poison; in many areas these activities are probably the most 'traditional' of farmers' responses to wildlife foraging on crops (Hill 1997; Naughton-Treves 2001; Hill 2004).

Methods to protect crops are most likely to be used, adapted over time and perceived to be effective if supported by farmers (Osborn and Parker 2003; Sitati and Walpole 2006). Farmers' capacity to protect their crops may involve compromises between desirable techniques and those that are feasible and affordable. These limitations can lead to frustrations for farmers that fuel 'conflicts' around wildlife rhetoric, especially when crop damage is persistent and involves considerable costs. Because farmers typically approach crop protection with few resources it is critical to understand wildlife foraging behaviour in order to maximize deterrent utility. The utility of an effective deterrent may be derived quantitatively by comparing crop-loss savings with costs of the deterrent over time. However, deterrent utility is also qualitative and will be influenced by a farmer's perceptions and expectations about the effort and opportunity costs to implement, use and maintain the technique (Conover 2002; Osborn and Parker 2003; Forthman, Strum and Muchemi 2005; Osborn and Hill 2005; Sitati and Walpole 2006; Graham and Ochieng 2008).

Farmers' attitudes and perceptions about crop damage and wildlife species causing this damage should be assessed when developing deterrent strategies, to ensure actions and anticipated outcomes are likely to address concerns (Pirta, Gadgil and Kharshikar 1997; Gillingham and Lee 1999; Osborn and Hill 2005; WWF 2005). Consultation with farmers and other key local people is imperative to agreeing plans and goals, and also staying informed about the steps, limitations, local resources or timeframes to take into account (Strum 1986; Osborn and Parker 2003; Hill 2004). Interventions should fit with local social norms, customs, or constraints, gender- or age-specific roles and labour availability over agricultural seasons (see Hill, Osborn and Plumptre 2002). Involving farmers actively in planning deterrent strategies often increases their ownership of techniques and commitment to finding solutions, reducing dependence on 'outsiders' (Hill, Osborn and Plumptre 2002; Osborn and Parker 2002a; Sitati, Walpole and Leader-Williams 2005; Graham and Ochieng 2008). There may be fewer calls for compensation for damage by wildlife when deterrents are developed with farmers and provide ongoing savings in crop loss, which is important because compensation does not reduce the incidence of crop damage, diminishes incentives to protect crops and can lead to deliberate crop damage and/or exaggerated claims (Hoare 1995; Rollins and Briggs 1996; Plumptre 2002; Bulte and Rondeau 2005; Sitati and Walpole 2006).

A deterrent used frequently, widely or over long periods of time is not necessarily effective; rather, it could be used because farmers lack, or perceive they lack, feasible alternatives. This is often the case for small-scale farmers, where traditional deterrents may be labour-intensive, inefficient and ineffective but the only options a farmer can afford and access with available resources (Hill, Osborn and Plumptre 2002). However, traditional deterrents such as basic fences, guarding and vigilance, or throwing objects, chasing and even culling animals feeding on crops may be appropriate in many contexts, especially when used systematically (see Osborn 2002). Traditional deterrents should be considered first when planning to address crop losses to animal foraging: techniques that farmers are familiar with are likely to be locally acceptable and readily modified to suit different circumstances, as well as easier to build on when greater efficiencies are required (Hill, Osborn and Plumptre 2002; Osborn and Hill 2005). The costs of deterrents and availability of resources must be factored into crop-protection strategies to ensure they retain value (i.e. are effective, efficient and affordable) for individual farmers over time.

In this chapter we describe the process of developing, implementing and evaluating a series of crop-protection tools in partnership with subsistence farmers in north-western Uganda. We reflect on the practicalities, benefits, potential costs and effectiveness of such an approach, and outline intervention considerations for future projects designing and implementing methods to deter crop-raiding.[2]

Case Study:
Partnering with Farmers in Uganda to Deter Crop Damage by Wildlife
Background

The study took place in six villages (Nyakafunjo, Nyabyeya 2, Kyempunu, Fundudolo, Marram, Panyana) around the southern edge of the Budongo Forest Reserve, Masindi District, Uganda (figure 9.1). Based on national census and population growth figures, there were approximately 5,300 people living in the study area on project commencement in 2006. Up to 70 per cent of people within the study area rely on agriculture for livelihood and subsistence farming predominates (Hill 1997; Webber 2006); consequently, any activity or event undermining crop yields will have potentially widespread impact on local food security and livelihoods. These can include stochastic weather events (such as onset of seasonal rain, storms, drought and lightning strikes igniting fires), crop production and market prices, seed quality, seed supply, soil fertility, crop damage by insects or disease and crop consumption by wildlife (Tweheyo, Hill and Obua 2005). Many local farmers perceive that wildlife foraging on crops is the problem with greatest risk experienced on their farm (Webber and Hill 2014).

The research was conducted in accordance with institutional ethics requirements and clearance, established guidelines for ethical social and primate research and with the consent and support of village councils and participating farmers. Permissions for the research were granted by the Uganda National Council for Science and Technology, Uganda Wildlife Authority, and National Forestry Authority, Uganda. The research occurred in the same area as earlier investigations of crop damage and farmer-wildlife interactions by Catherine Hill and Amanda Webber. Working in the same villages, and with some of the same farmers as Hill and Webber, provided context and continuity for the study. Additionally, the goodwill generated in local communities by the previous research helped to ensure local support for the project. This positive foundation at commencement of the study meant that introductions and logistics within study villages were considerably more straightforward than they otherwise might have been. Chairpersons of each study village were provided with details of the proposed project, objectives, methods, and timetable and then requested to permit the research to be conducted in the village and also allow GW

Figure 9.1. Map showing the location of Budongo Forest Reserve in north-western Uganda. *Source:* NordNordWest/Wikipedia; Wikimedia Commons.

to approach individual farmers for recruitment. Permissions were granted in all cases. Farms were identified for inclusion in the study after evaluating their extent of view of forest edges, history of crop damage by wildlife and planned crops (Hill 2000; Webber 2006); thirteen farms were selected and each was located at the forest edge (see Wallace and Hill 2012). Predominant crops on study farms, as well as across the study area, were maize (*Zea mays*), beans (*Phaseolus vulgaris*), sorghum (*Sorghum bicolor*), cassava (*Manihot esculenta* and *Manihot palmata*), millet (*Eleusine coracana*) and bananas (*Musa* spp.). Key project stakeholders were farmers and their families, others supported by farming, local communities, village councils and government agencies; other interested groups were a local association of sugar cane farmers as well as non-governmental organizations.

Project Design

The original research was designed to develop, implement and evaluate a set of effective and locally acceptable crop-protection tools to reduce the impact of wildlife on subsistence livelihoods in farming areas adjacent to forest habitat (Wallace and Hill 2012). Including local farmers throughout all stages of the project was of primary importance and this is reflected in the research design (Hill and Wallace 2012). An interview averaging forty-five minutes in duration was conducted with each study farmer (1) prior to commencing systematic on-farm observations, (2) at the end of Year One and (3) at the end of Year Two. Three 3-hour focus-group sessions, as well as an end-of-study results and training workshop, were also conducted with farmers and village chairpersons. Each focus group provided a formalized opportunity for farmers to contribute to developing deterrents and reflect on the process.

The aim in Year One was to acquire baseline information about farmer and wildlife behaviour around crop-foraging events. All study farmers were encouraged to carry out their usual farming practices and patterns of activity, including responses to animals foraging on crops, vigilance and deterrent behaviour. A crop-raiding event (CRE)[3] was defined to occur when one or more individuals of an animal species entered a farm (i.e. crossed a farm boundary), interacted with one or more crop stems and left the farm (Wallace 2010). The CRE commenced when the first individual entered the farm and ended when the last individual exited. Baseline data on the behaviour of people and animals (wild and domestic) were systematically recorded to (1) identify species responsible for damaging crops, (2) establish the frequency and extent of crop-damage events, (3) examine farmer detection of and responses to animals on or around their farms and (4) explore animal responses to existing strategies for crop protection utilised by farmers. The results of these initial observations were fed back to farmers through individual discussions at the end of Year One, including a written summary of the key findings for their farm, as well as focus-group sessions prior to Year Two. Although it was originally intended to record crop-foraging events during the day and night, local logistics and the need to utilize artificial light sources to observe nocturnal species precluded data collection at night and the study focused on diurnal species. However, many farmers reported crop-foraging by nocturnal animals, mainly bush pigs (*Potamochoerus porcus*) and porcupines (*Hystrix cristata*), and some deterrents were developed to address this.

Focus-group sessions were conducted with farmers and village chairpersons prior to Year Two to explore options for deterrents, considerations when constructing and monitoring deterrents and prior experiences with crop-protection techniques. Each session was in a different village and farm to minimize travel distances for the participants. A broad range of ideas and options for deterrents were generated by farmers; most were uncomplicated and affordable, and could be readily implemented. Farmers were also positive about potential deterrent techniques suggested by the research team from analysis of crop-foraging activity during Year One, and did not deem any proposed options unacceptable. The initial behavioural observation results, in conjunction with ideas generated by focus-group participants, were used to identify the primary aspects of farmer-animal interactions that could be manipulated to reduce crop damage by animals: these were improving farmer detection of animals entering farms and reducing wild animals' ability and/or willingness to enter crop fields.

In contrast, for Year Two the aim was to assess the efficacy of deterrents. Consequently, study farmers were encouraged to modify their crop-protection efforts or adopt new methods. This often involved farmers adjusting their behaviour according to the perceived benefits, demands or shortcomings of each deterrent implemented at their farm. Deterrents were tested on thirteen farms across the six study villages. Each of the farms had at least one boundary adjoining forest that provided natural habitat for primates and other wildlife; each of these boundaries comprised a farm-forest interface (FFI).

As agreed with farmers prior to onset of the research, researchers had an unobtrusive and passive role on farms, not responding to animals entering farms and also not disclosing wildlife presence or foraging activity to any people on farms. Study farmers also actively participated in data collection by reporting a broad range of details for each crop-damage event they detected (and usually responded to) when the research team was not observing at their farm. These damage events (referred to as reported CREs) were additional to those observed by the research team during systematic sampling (referred to as observed CREs). Information provided by farmers was compiled regularly and at least twice each week. As far as possible, parameters recorded for each reported CRE were the same as for observed CREs provided the event could be substantiated. Substantiation was via signs of wildlife, such as tracks or scat, and evidence of crop damage.

Farmers were advised before observational sampling commenced that all details about crop-damage events that occurred when researchers were not at their farm would be helpful and informative for the research. While farmers did not routinely describe crop damage in systematic terms, it became evident that they could recall accurately a broad range of crop-damage event parameters when guided about which were relevant, particularly those directly influencing the amount of stem damage sustained (Wallace 2010). Recall reliability was assessed regularly with each farmer by having them describe a CRE they had responded to during an observation session and then comparing the details recalled with researcher records. In each case the crop-damage event had occurred at least one day prior to recall assessment to simulate the average time interval between when damage by wildlife occurred and when the CRE was reported to the research team. Each farmer was able to recall the key parameters of the nominated CRE at almost 90 per cent concordance with researchers' records. Farmers could identify reliably each animal species, including primates, commonly occurring near their farm. Interestingly, farmers appeared to be interested mainly in recalling details of crop damage by wildlife, even though advised that damage to crops by domestic farm animals, such as pigs or goats, was equally important to the study.

Observing and Understanding the Linkages between Farmer and Wildlife Behaviour

Primates were involved in 96 per cent of observed CREs by wildlife (n = 227), accounting for 99 per cent of crop stems damaged (n = 4,168). Species observed damaging crops were olive baboons (*Papio anubis*), red-tailed monkeys (*Cercopithecus ascanius schmidti*), vervet monkeys (*Chlorocebus aethiops*), blue monkeys (*Cercopithecus mitis stuhlmanni*), chimpanzees (*Pan troglodytes schweinfurthii*) and black and white colobus monkeys (*Colobus guereza occidentalis*). Maize and beans were the crops foraged on most frequently across study farms (Wallace and Hill 2012).

Farmers' views of disturbance caused by primates and other wildlife, gauged from responses during semi-structured interviews, were very similar across the sample and broadly concurred with our independent observations summarized above. All farmers stated that primates cause more crop damage than other wildlife, although most added that bush pigs can damage crops extensively. Baboons were ranked by farmers as the worst animal for causing crop damage, bush pigs were ranked second, and monkeys other than baboons were ranked third. All farmers stated that maize was damaged most often and extensively by primates, followed by beans, bananas or millet.

Observations of farmer behaviour, monitoring of farms and discussions with farmers confirmed they use various, although not necessarily effective, methods (such as scarecrows, village bells, sporadic guarding, or vine fences) to protect crops but are aware of more potentially effective methods (including alarm systems, wire fences, and extensive guarding) they can feasibly access. It also became apparent that many farmers did not use crop-protection techniques to the extent they said they did. A striking example of variation between stated behaviour and observed behaviour was guarding activity. Discussions with farmers confirmed that they (and the research team) considered guarding to involve presence in crop fields near forest edges for extended periods of time or continuously, patrolling crop fields, vigilance and scanning for wildlife approaching or entering the farm and active responses if animals are detected; most farmers also considered this the ideal situation, which was sometimes not possible.

Patterns of guarding described by farmers were markedly different to the patterns observed in Year One. Far from occurring throughout each day over the season, guarding comprised only 15.5 per cent of the farmer activity budget across all study farms (Wallace 2010), yet farmers remarked that if they did not 'guard well' (i.e. diligently) at the times they said they guarded then they would lose from one to three quarters of their crop stems to wildlife. Farmers linked amounts of loss directly to the extent and quality of guarding. Most farmers (76.9 per cent) reported guarding at their farm was organised, intensive (i.e. approached seriously and with a large time investment, often at the expense of other activities), involved all family members except infants, and occurred in shifts to ensure crops were always protected. When interviewed at the end of Year One, knowing they had been observed, 42 per cent of farmers stated they had guarded to a greater extent than in previous seasons while

58 per cent 'admitted' they had done little guarding over the season compared to the amount originally stated. Similarly, six of eight farmers who stated they guarded for most of each night probably did not do so. Although ten farmers claimed to light fires at night near a guard hut or the farm-forest edge to signal their presence to bush pigs, there were signs of fires at only two farms where some guarding at night probably occurred.

Overall, the extent of guarding that occurred at each study farm during Year One was probably too limited and sporadic to be effective for crop protection. In many instances, any value guarding had as a deterrent would have been compromised by poor quality of guarding behaviour.[4] Very few individuals observed guarding were vigilant or patrolled fields for longer than one bout of approximately thirty to sixty minutes. Farmers appear to consider guarding as a sedentary rather than dynamic activity. Children were distracted regularly while guarding and were often observed to play instead, even occasionally breaking maize stems while playing.

Most guarding activity appeared to be carried out by adult women, which is common within subsistence farming communities (see Colfer, Wadley and Venkateswarlu 1999; Hill, Osborn and Plumptre 2002; Webber 2006). This could have reduced the time women allocated to household tasks and/or alternative sources of income; however, activity budgets confirmed it was not directly due to men allocating most of their time to working on-farm. At several study farms men were absent more than women and more likely to participate in logging or pit-sawing within the forest. Overall, the labour and/or opportunity costs of guarding (Naughton-Treves 1997; Hill, Osborn and Plumptre 2002; Gillingham and Lee 2003) were relatively low at study farms because guarding did not comprise a large proportion of farmers' time and did not occur at expense of work to grow food or generate income.

Using systematic observational techniques in partnership with study farmers made it possible to determine that they did relatively little to actively or passively protect their crops from wildlife, contrary to their claims, as also found by Liva Hansen (2003) about farmers in Sulawesi, Indonesia. Systematic observations also revealed that, in most cases, techniques of crop protection used by farmers (such as scarecrows, intermittent guarding, and vine fences) were ineffective, or only sporadically effective. Although this was probably partly due to inconsistent crop-protection behaviour, some farmers indicated they use these methods because they do not know what else to do to deter animals and affordable alternatives are lacking.

Approach and Rationale when Developing Deterrents

It was important that the deterrents introduced at study farms were easy for farmers to use, relatively or potentially affordable, used materials farmers could source locally, and required minimal maintenance in terms of effort and cost; otherwise the methods would be impractical and of little value for farmers. Within this context, however, the primary objectives for the research were to (1) determine which deterrent methods reduced crop-foraging, (2) assess how the deterrents operated, (3) establish why they were effective or ineffective and (4) gauge how useful the techniques could be for farmers. These aims meant that when alternative crops were trialled as deterrents such as ocimum, *Ocimum kilimandscharicum,* and jatropha, *Jatropha curcas*, cash saleability or immediate profitability of the crop for a farmer was not the driving factor in deterrent evaluation. Because economic conditions and markets for crops will almost invariably fluctuate over time (Bigagambah 1996; Padulosi, Eyzaquirre and Hodgkin 1999), developing deterrents that were effective in terms of impacts on crop-foraging took precedence over short-term crop marketability. It was also anticipated that donor development and conservation agencies would be more likely to help farmers defray the costs of deterrents when efficacy was demonstrated and measurable.

Reduced crop loss was the main measure of deterrent value for farmers but they also identified long-term effectiveness as an important consideration when deciding to invest in a deterrent. Because farmers viewed deterrent value in terms of costs and benefits over time, it was possible to evaluate methods that used relatively long-lasting but more-expensive materials that a local farmer would need to plan to purchase from savings rather than access from the forest. It was important that forest resources were not essential to construct and/or operate the deterrents.

Implementing, Monitoring and Evaluating Deterrents

Baseline results were used to assess which deterrents would be (1) best evaluated and rigorously tested at each FFI and (2) of most benefit to individual farmers given the frequency and patterns of crop loss experienced. This information was used in conjunction with each farmer's preferences to allocate deterrents for evaluation. Farmers had input into which deterrents were trialled at their farm, and were able to veto any suggestions they disagreed with or did not consider practical. Ongoing consultation with farmers ensured each deterrent was introduced and monitored in partnership with them. It was agreed with farmers that it was not feasible to test all deterrents at each farm. Rather, it was acknowledged that the aim of the intervention stage of research was to determine which deterrent methods were effective, which were not effective, and why, and this would be best achieved by matching deterrents to specific FFIs.

Farmers were advised that deterrent techniques can differ in impact at different farms and also with different animal species. Therefore, their experiences of a deterrent might vary across the season as part of determining whether and how the technique worked, and whether it operated specifically or broadly. Because most study farmers knew each other and shared experiences, it was essential to advise them from the outset of Year Two that some might have better results than others from similar deterrents (as well as different deterrents) and this was an important aspect of deterrent evaluation. Otherwise, some might have been disappointed or disillusioned if they perceived they were not receiving the same benefits as others. Farmers were assured the results and information gained from testing each deterrent would be shared with all of them on project completion, and would therefore benefit each farmer in subsequent seasons. Farmers were also advised that it could be necessary to adjust or modify techniques over time because animals often habituate to deterrents and develop ways to bypass them. Primates are especially likely to do this due to their intelligence, behavioural flexibility, dexterity, cooperative capacity and social organization (Maples et al. 1976; Strum 1994; Hill, Osborn and Plumptre 2002; Forthman, Strum and Muchemi 2005; Osborn and Hill 2005).

An extensive range of tools, techniques and structures to deter crop-foraging wildlife were implemented at study farms in Year Two, including barbed-wire fences, mesh fences, rope fences, solar lights, trenches, vegetation barriers, chilli powder, nets, bells, systematic guarding, dogs, glasses for farmers, alternate locations and timing for crops and synchronization of crops. Some of the deterrents were used in combination (table 9.1). We could not guarantee outcomes for farmers, therefore all costs for materials and related expenses (including labour to transport materials to farms) were paid from project funds. Construction of deterrents was undertaken by the research team together with farmers. Although each farmer incurred effort costs in trialling a deterrent, project funding ensured any benefits accrued without monetary investment, allowing farmers to focus on deterrent utility and effectiveness. It is believed that being provided with deterrents did not bias farmers' use of any deterrents or perceptions about them (Heong and Escalada 1998; Bernard 2002). Farmers were assured deterrents could be removed at their request at any time, and would be removed on completion of the study unless they requested otherwise.

Several deterrents at study farms improved farmer detection of CREs, namely paid guards, net fences with bells and alarms generally (see table 9.1 for a summary); further details are reported by Hill and Wallace (2012). Farmers may also have been more focused on being vigilant during Year Two because of deterrent-related activities. Farmers were aware of the Year One results prior to Year Two, which could have modified their vigilance in a manner similar to epistemic feedback noted in organizational behaviour research (Rollinson 2005; Buchanan and Huczynski 2010). Compared to Year One baseline values, several farmers were observed to modify their behaviour during Year Two as a result of deterrent implementation. In most cases farmers allocated less time to guarding because they perceived deterrents reduced the need to guard, even where it could have enhanced deterrent efficacy (for example, in conjunction with vegetation barriers). Farmers who decreased their guarding effort allocated more time to work and income activities such as weeding crops, brewing and pit-sawing. Although these shifts in activity were not anticipated, they provide insights about the potential opportunity costs of protecting crops.

Each farmer reported the deterrents trialled at their farm to be useful and effective, and all stated they gained benefits from the deterrents and would continue to use them. Accordingly, each farmer requested that deterrents remain *in situ* for ongoing use or extension. Farmers' favourable ratings for deterrents could have reflected an intent to please the research team and provide 'correct answers' (Morton-Williams 1993; Esterberg 2002; Randall and Koppenhaver 2004), especially because they were funded by the project, but it was evident they perceived the deterrents as valuable and responsible for reduced crop loss. From the farmers' perspective the deterrents mitigated crop losses to wildlife, and to a large extent this was probably achieved

Table 9.1. Deterrents trialled, their capacity to reduce crop damage by animals, animal species deterred and farmers' evaluations.
Source: Adapted from Hill and Wallace (2012) and Hsiao et al. (2013).

Deterrent	Capacity to reduce crop damage	Animal species deterred effectively	Farmer evaluation Benefits	Farmer evaluation Disadvantages
Paid guards	Yes.	Primates.	Effective when systematic. Saves farmer's time.	Expensive.
Guard dogs	No, except as a form of alarm.		Able to detect raiding animals.	Costs to feed and vaccinate dogs.
Barbed-wire fence	Yes.	Bush pigs. Larger primates, especially if linked with active guarding or chasing.	Effective and long-lasting. No need for farmer to be present.	Expensive to construct and also to replace when required.
Net fence with bells	Yes.	Primates, especially smaller species.	Allows farmer to work away from farm edges.	Requires farmers to be present to respond to raids.
Rope fence with bells	No. Potentially, if constructed to border all fields.		Potentially useful alarm if borders many crop fields.	Also requires active guarding.
Rope fence with chilli powder	Yes.	Primates, especially smaller species.	Effective and does not require farmer presence.	Cost of chilli and time to prepare.
Ocimum hedge	Yes, when used with wire mesh fence.	Primates.	Hardy, easily grown as a dense hedge. Potential cash crop.	Lifespan of three years then need to replant. Uses land for crops.
Jatropha hedge	Yes.	Primates. Bush pigs.	Hardy, inexpensive, easy to propagate and extend. Other household fuel uses.	Lead time needed to establish as a hedge. Can be prone to damage by borers.
Trench	No.		Usually inexpensive.	Must keep clear.
Alarms generally	Yes.	Primates.	Alerts farmer before crops are damaged.	Farmer must be present and respond.
Solar lights	Yes.	Bush pigs.	Effective and do not require fuel. Other household uses out of crop season.	Costs to replace. Easily stolen. Need to be recharged in sunlight daily.

because farmers could 'buy in' via involvement at all stages of the intervention process. Many of the deterrents were still in use a year after the research was completed and numerous farmers not included in the original study had installed similar or modified versions of the trialled deterrents on their own farms (Hsiao et al. 2013).

Farmers rated deterrents on a cost-benefit basis. Although some deterrents (such as extensive barbed-wire fences, mesh fences, solar lights or full-time systematic guarding) were not readily affordable for many farmers, it was generally acknowledged that savings in crop losses and labour would usually offset costs over time. In addition, demonstrating and measuring the effectiveness of these and similar deterrents probably increased their suitability for support from conservation and development agencies (see O'Connell-Rodwell et al. 2000; Sitati, Walpole and Leader-Williams 2005). Deterrents such as jatropha, ocimum, and chilli were perceived as very cost effective because they can be readily grown by farmers and require minimal maintenance (table 9.1).

When interviewed at the end of Year Two, one farmer confirmed barbed-wire fences 'work very well for stopping bush pigs'; his family was 'very happy to have the fence' because he could get more sleep at night and they lost less maize than in other seasons. The farmer nominated additional sleep due to less guarding at night as a major benefit of the fence because he could then work on the farm for 'many more hours each day for most of the season'. As a result, it had not been necessary to pay people to help weed his crop fields, as required when he spent more time guarding at night. Additional time to work on the farm had also allowed his wife to spend more time on household tasks. The farmer envisaged the fence would require only minor maintenance for three to four years, and the barbed wire would outlast the posts due to termite damage. Although the cost of replacing the wire was viewed as a potential shortcoming of the fence, the farmer added he had 'many years to save the shillings' for this and the fence would 'help him to save more money' over future years (Wallace 2010).

Another farmer stated solar lights were 'powerful and could be seen from many farms away' and were 'very good for scaring or stopping bush pigs'. The farmer envisaged the lights would also be very useful as lighting in his house between growing seasons, saving him the cost of other fuels. He added that solar lights were easy to recharge and use, and flexible deterrents because they could be moved to other locations as required; however, because the lights were useful and conspicuous, the farmer felt they were at high risk of being stolen. The farmer ensured the lights were collected from the farm edge at dawn, recharged at his house each day, and many people in the village knew the lights were his property (Wallace 2010).

A farmer trialling chilli paste applied to a rope fence stated that chilli paste was 'easy to mix' and 'powerful', and key to the fence's effectiveness. The farmer added that almost one hour was required every three to four days to apply the paste to the fence, which was a small time investment relative to benefits. The farmer planned to maintain the fence, replace the sisal rope when necessary and grow chilli for paste (compare with Graham and Ochieng 2008). Although the farmer noted the barrier may be less useful in periods of frequent or heavy rain because this would remove the chilli too often or weaken the sisal, he acknowledged such rain usually only occurred early in each season before crops were foraged on regularly (Hill and Wallace 2012).

A farmer growing jatropha as a barrier stated he intended to extend the fence around his farm because it was effective, required very little maintenance, and was cost free. The fence was not damaged by animals, did not use materials needing regular replacement and required infrequent pruning to maintain density. Another farmer identified two potential shortcomings of growing ocimum as a deterrent. Firstly, because plants have a relatively short lifespan of three to four years it is necessary to supplement the hedge with new plants every few years, and therefore he would need to set aside land for seed beds and spend time growing seedlings. Secondly, over time an ocimum hedge could occupy two or more metres of land at the farm edge, rendering that area unavailable for growing staple crops. However, the farmer added that ocimum could also be a viable cash crop if local or regional marketing networks were created and used (see Barbier 1987; Bigagambah 1996; Padulosi, Eyzaquirre and Hodgkin 1999). A farmer testing a single-row ocimum hedge noted that it occupied very little land and considered this to be a key benefit, particularly as ocimum retained animal-repellent properties while planted in one row; although the fence was difficult for baboons and (other) monkeys to jump or climb over, its efficacy was also tied to the odour of the ocimum and presence of bees. The farmer also agreed ocimum was potentially valuable as a cash crop, but noted it would be important to have a large group of farmers or a commercial organization promoting it.

It is recommended elsewhere that neighbouring farmers should assist each other to chase crop-foraging wildlife from their farms (see Hill, Osborn and Plumptre 2002), and guarding could be most effective and cost efficient when conducted cooperatively over several farms or shared with a neighbour (Maples et al. 1976; Warren, Buba and Ross 2007). However, only two farmers in our study stated they shared or would consider sharing their resources to protect crops, even though there were a number of observations of farmers assisting neighbours to chase animals from their farm. Some farmers viewed other people as lazy or careless about crop protection and not using an equal or appropriate amount of effort when guarding the crops of others; they therefore considered it preferable for farmers to focus on protecting their own crops, which typically leaves little time to assist others (H. Biroch pers. comm.).

Deterrent effectiveness was often greater when used in combination, such as ocimum with a mesh fence, jatropha with strands of barbed wire, alarm systems with rapid responses to animals entering crop fields, and barriers patrolled by guards. Most deterrents require association with guarding and responses by farmers to maintain efficacy over time; otherwise raiders will perceive deterrents to rarely carry aversive consequences. Barbed-wire fences deterred primates most when linked with farmers' responses (such as chasing or throwing objects), demonstrating that active protection is preferable to a passive barrier, as also noted by others (Osborn and Parker 2002b; Sitati and Walpole 2006). While combining techniques might be a practical way to maintain deterrent novelty and minimize habituation, it will be prudent to consider costs to farmers. For example, wire fences are sizeable investments, and it will be important to determine whether such fences add protective value to vegetation barriers that are readily grown and maintained at low cost. Similarly, guarding may involve opportunity and social costs for farmers, and should be balanced with deterrent return.

Reflecting on the Importance of Engaging with Local Priorities, Perceptions and Realities

Interviews and focus groups with farmers complemented systematic observations of wildlife and farmer behaviour to develop locally appropriate and affordable deterrents to raiding. Drawing upon baseline data from an entire growing season, it was possible to evaluate the efficacy of a range of crop-protection techniques, under almost experimental conditions, over the subsequent season. As demonstrated here, farmer-reported records provide useful information about wildlife crop-foraging activity, complement data from systematic observations by researchers and are sufficiently reliable for analysis when potential limitations are acknowledged. Most importantly, the farmers' role in data collection promoted their ownership and involvement in the project, and their commitment to participatory solutions. The observation that farmers were most interested in recalling crop foraging events by wildlife rather than domestic farm animals probably reflects commonly reported differences in how these groups of animals may be viewed. Farmers frequently tolerate even relatively high levels of crop damage by farm animals because they are financial assets and/or potential sources of food, and farmers often have protocols for compensating each other if their animals damage crops (Naughton-Treves 1998; Hill 2004).

Local people commented that research studies, development projects or tourism ventures were often conducted within the area, without seeking their views and involvement, or providing community benefits (also see Lauridsen 1999). Not all local people are aggrieved in these ways and many have positive perceptions of 'outsiders'; however, discussions in villages confirmed that a large number of people may take advantage of the naiveté and resources of outsiders if they perceive this to be possible. This can bias the verbal information collected in field studies (Brislin, Lonner and Thorndike 1973; Bernard 2002), and highlights the importance of understanding local context as well as building trust, communication and local participation into project planning.

Because the deterrents were developed in partnership with farmers, they were deemed locally appropriate and acceptable, even if unusual in the study area: for example, solar lights. While similar research has noted that farmers often fail to suggest crop-protection methods or appear disinterested and dependent on ideas from others (see Hill, Osborne and Plumptre 2002; Warren 2003; Webber 2006), this was not found in the present study. Farmers were enthusiastic and actively involved in assessing deterrents, often exchanging 'progress reports' about *their* deterrents' utility. Other local farmers did not ostracise study farmers for having fully-funded deterrents at their farms, which is not always the case (see Hill, Osborn and Plumptre 2002). Rather, other villagers supported the project because they envisaged benefits for many farmers over time (M. Diedonne, A.H. Fani, S. Oliki, and M. Kakole pers. comm.). A training and feedback workshop was conducted with all study farmers and village chairpersons at the end of Year Two to ensure effective dissemination of project results and information about how to install the deterrents. Farmers were encouraged to visit study farms to observe the range of deterrents used; study farmers also confirmed they would share deterrent information within their village. It is believed this participatory and cooperative approach was central to farmers' positive perceptions of deterrents (see Osborn 2002; Osborn and Hill 2005; Sitati, Walpole and Leader-Williams 2005; Sitati and Walpole 2006). Although demonstrating deterrent effectiveness does not necessarily ensure use by farmers (Sitati and Walpole 2006; Graham and Ochieng 2008), positive perceptions increased the likelihood that farmers would maintain and adapt the deterrents, and continue to view them favourably (Hsiao et al. 2013).

Primates probably adjust their crop-foraging behaviour to avoid detection by farmers and bypass alarm systems, and are unlikely to be deterred easily if crops benefit them. Consequently, any techniques used to improve farmers' detection of crop foraging will require adjustment over time to remain effective. The influence of specific sets of factors on detection might also be dynamic, changing as farmers and primates extend their experience of crop-foraging. There is no reason to presume that factors influencing animal crop damage at a farm today will always do so, or will to the same degree. Accordingly, crop-protection methods must be modifiable for changes in conditions as well as adaptation by wildlife or farmers.

Conclusion

The study not only demonstrates that it is possible to develop, implement and monitor a range of effective deterrents to reduce crop-foraging, but also confirms the importance of doing this in partnership with farmers. Participatory projects often build relationships between stakeholders, improve communication, promote better understanding of alternative perspectives and reduce conflict and misunderstandings (Newmark et al. 1993; Hulme and Murphree 2001; Madden 2004). However, while learning and working together with farmers may be beneficial in itself, ineffective or misguided interventions that waste time and resources can undermine interest in mitigation goals (see Stewart, Coles and Pullin 2005). Deterrents should also not be viewed as an entire or 'set and forget' solution to crop damage issues; they may simply address symptoms rather than causes (Barnes 2002). The greater the extent to which intervention strategies (1) are based on a thorough and accurate account of crop foraging activity, (2) incorporate an understanding of crop-foraging animal and farmer behaviour, and (3) involve farmers and acknowledge their interests and perceptions, the greater the probability of those strategies being effective. Before implementing an intervention to protect crops it is imperative to understand the context of crop damage by wildlife. Because human dimensions are central to this context, it is critical to work alongside stakeholders as partners when developing and implementing intervention plans.

Acknowledgements

For permissions to conduct the research we thank the Uganda National Council for Science and Technology, Uganda Wildlife Authority, and National Forestry Authority, Uganda. Financial support was provided by The Leverhulme Trust, American Society of Primatologists, North of England Zoological Society, Rufford Small Grants, Parkes Foundation, and Primate Society of Great Britain. Mawa Diedonne, Andama Hanington Fani, Amandu Geoffrey, Johnson Ayebale, Helen Biroch, Alfred Awio, Phillip Madrira, Enzama Anthony, Rob Bacon, Tim Pearson, and Andrea Wallace provided invaluable field assistance. We especially thank the farmers, families, and village chairpersons of Nyabyeya Parish for their key involvement, support, enthusiasm, and hospitality throughout the research.

Catherine Hill is Professor of Anthropology in the Department of Social Sciences at Oxford Brookes University. Her research focuses on people-wildlife relationships, conservation conflicts and implications for people's perceptions of wildlife, biodiversity conservation and local communities.

Graham Wallace obtained a PhD at Oxford Brookes University and an MBA through Durham University Business School. With a focus on applied research, his interests include wildlife ecology, interactions between wildlife and humans, behavioural economics in conservation, anthropogenic impacts on ecosystems and implications for management strategy. Graham has extensive experience working with rural communities in sub-Saharan Africa, and is currently engaged in coastal marine ecology research in British Columbia, Canada.

Notes

1. Vertebrate species reported to consume crops and other human foods are listed in Hygnstrom, Timm and Larson (1994), Naughton-Treves (1997: 32), Conover (2002), CARE et al. (2003), Warren (2003: 4, 8), Lee and Priston (2005: 4–8), Hill (2005: 43), Webber (2006: 9–10, 13–15) and Priston and McLennan (2013).
2. The term 'deterrent' refers here to any technique intended to protect crops from damage by animals and can be a sensory stimulus, structure, tool, action or procedure. The term therefore encompasses the broad general range of alarms, repellents, barriers, alternate crop locations and synchronised planting, as well as traditional forms such as guarding, fires or scarecrows (Hill, Osborn and Plumptre 2002; CARE et al. 2003; Lee and Priston 2005; Osborn and Hill 2005; Sitati and Walpole 2006).
3. Here we retain the term CRE (crop-raiding event) used in earlier publications to facilitate comparison; however, we recommend this term be revised to 'crop-foraging event' in future publications.
4. This would mainly comprise lack of vigilance and/or not being ready to respond to wildlife incursions.

References

Abbot, J.I.O., et al. 2001. 'Understanding the Links between Conservation and Development in the Bamenda Highlands, Cameroon', *World Development* 29(7): 1115–36.
Adams, W.M. 1998. 'Conservation and Development', in W.J. Sutherland (ed.), *Conservation Science and Action*. Oxford: Blackwell Science, pp. 286–315.
Altrichter, M. 2006. 'Wildlife in the Life of Local People of the Semi-Arid Argentine Chaco', *Biodiversity and Conservation* 15(8): 2719–2736.
Armitage, D.R. 2003. 'Traditional Agroecological Knowledge, Adaptive Management and the Socio-politics of Conservation in Central Sulawesi, Indonesia', *Environmental Conservation* 30(1): 79–90.
Baker, J., et al. 2013. 'Linking Protected Area Conservation with Poverty Alleviation in Uganda: Integrated Conservation and Development at Bwindi Impenetrable National Park', in J.B. Smith (ed.), *National Parks*. New York: Nova Science Publishers, pp. 47–103.
Barbier, E.B. 1987. *Cash Crops, Food Crops and Agricultural Sustainability*. Gatekeeper Series No. 2: International Institute for Environment and Development, Sustainable Agriculture and Rural Livelihoods Programme.
Barnes, R.F.W. 1996. 'The Conflict between Humans and Elephants in the Central African Forests', *Mammal Review* 26(2/3): 67–80.
Barnes, R.F.W. 2002. 'Treating Crop-Raiding Elephants with Aspirin', *Pachyderm* 33: 96–99.
Beale, C.M., and P. Monaghan. 2004. 'Behavioural Responses to Human Disturbance: A Matter of Choice?' *Animal Behaviour* 68(5): 1065–1069.
Bernard, H.R. 2002. *Research Methods in Anthropology: Qualitative and Quantitative Methods*, 3rd ed. Walnut Creek, CA: AltaMira Press.
Bigagambah, J.R. 1996. *Marketing of Smallholder Crops in Uganda*. Kampala, Uganda: Fountain Publishers.
Blomley, T. 2000. *Woodlots, Woodfuel and Wildlife: Lessons from Queen Elizabeth National Park, Uganda*. Gatekeeper Series No. 90: International Institute for Environment and Development, Sustainable Agriculture and Rural Livelihoods Programme.
Boydston, E.E., et al. 2003. 'Altered Behaviour in Spotted Hyenas Associated with Increased Human Activity', *Animal Conservation* 6: 207–219.
Brislin, R.W., W.J. Lonner and R.M. Thorndike. 1973. *Cross-Cultural Research Methods*. New York: Wiley.
Buchanan, D.A., and A.A. Huczynski. 2010. *Organizational Behaviour*, 7th ed. Harlow: Pearson Education.
Bulte, E.H., and D.Rondeau . 2005. 'Why Compensating Wildlife Damages may be Bad for Conservation', *Journal of Wildlife Management* 69(1): 14–19.
Bush, G., et al. 2004. 'The Value of Uganda's Forests: A Livelihoods and Ecosystems Approach', *WCS Albertine Rift Programme, EU Forest Resources Management and Conservation Programme, National Forest Authority*, pp. 101.
CARE et al. 2003. 'Reducing the Costs of Conservation to Frontline Communities in Southwest Uganda. Knowledge Base Review Report', *CARE International in Uganda, Institute of Tropical Forest Conservation, Conservation Development Centre, and Wildlife Conservation Society*.
Chase, L.C., D.J. Decker and T.B. Lauber. 2004. 'Public Participation in Wildlife Management: What do Stakeholders Want', *Society and Natural Resources* 17(7): 629–639.
Chiyo, P.I., et al. 2005. 'Temporal Patterns of Crop Raiding by Elephants: A Response to Changes in Forage Quality or Crop Availability?' *African Journal of Ecology* 43(1): 48–55.

Choudhury, A. 2004. 'Human-Elephant Conflicts in Northeast India', *Human Dimensions of Wildlife* 9(4): 261–270.

Colfer, C.J.P., R.L. Wadley and P. Venkateswarlu. 1999. 'Understanding Local People's Use of Time: A Pre-condition for Good Co-management', *Environmental Conservation* 26(1): 41–52.

Conover, M. 2002. *Resolving Human-Wildlife Conflicts: The Science of Wildlife Damage Management*. Boca Raton, FL: Lewis Publishers.

Dickman, A.J. 2010. 'Complexities of Conflict: The Importance of Considering Social Factors for Effectively Resolving Human-Wildlife Conflict', *Animal Conservation* 13(5): 458–466.

Esterberg, K.G. 2002. *Qualitative Methods in Social Research*. Boston, MA: McGraw Hill.

Forthman, D.L., S.C. Strum and G.M. Muchemi. 2005. 'Applied Conditioned Taste Aversion and the Management and Conservation of Crop-Raiding Primates', in J. D. Paterson and J. Wallis (eds), *Commensalism and Conflict: The Human-Primate Interface*. Norman, OK: American Society of Primatologists, pp. 420–443.

Fuentes, A., M. Southern and K.G. Suaryana. 2005. 'Monkey Forests and Human Landscapes: Is Extensive Sympatry Sustainable for *Homo sapiens* and *Macaca fascicularis* on Bali?' in J.D. Paterson and J. Wallis (eds), *Commensalism and Conflict: The Human-Primate Interface*. Norman, OK: American Society of Primatologists, pp. 168–195.

Gillingham, S., and P.C. Lee. 1999. 'The Impact of Wildlife-Related Benefits on the Conservation Attitudes of Local People around the Selous Game Reserve, Tanzania', *Environmental Conservation* 26(3): 218–228.

———. 2003. 'People and Protected Areas: A Study of Local Perceptions of Wildlife Crop-Damage Conflict in an Area Bordering the Selous Game Reserve, Tanzania', *Oryx* 37(3): 316–325.

Graham, M.D., and T. Ochieng. 2008. 'Uptake and Performance of Farm-Based Measures for Reducing Crop Raiding by Elephants *Loxodonta africana* among Smallholder Farms in Laikipia District, Kenya', *Oryx* 42(1): 76–82.

Hackel, J.D. 1999. 'Community Conservation and the Future of Africa's Wildlife', *Conservation Biology* 13(4): 726–734.

Hansen, L.K. 2003. 'Influence of Forest-Farm Boundaries and Human Activity on Raiding by the Buton macaque (*Macaca ochreata brunnescens*).' MSc dissertation, Oxford Brookes University, Oxford, pp. 46.

Happold, D.C.D. 1995. 'The Interactions between Humans and Mammals in Africa in Relation to Conservation: A Review', *Biodiversity and Conservation* 4(4): 395–414.

Heong, K.L., and M.M. Escalada. 1998. 'Changing Rice Farmers' Pest Management Practices through Participation in a Small-Scale Experiment', *International Journal of Pest Management* 44(4): 191–97.

Hill, C.M. 1997. 'Crop-Raiding by Wild Vertebrates: The Farmer's Perspective in an Agricultural Community in Western Uganda', *International Journal of Pest Management* 43(1): 77–84.

———. 2000. 'Conflict of Interest between People and Baboons: Crop Raiding in Uganda', *International Journal of Primatology* 21(2): 299–315.

———. 2002a. 'People, Crops and Wildlife: A Conflict of Interests', in C.M Hill, F.V. Osborn and A.J. Plumptre (eds), *Human-Wildlife Conflict: Identifying the Problem and Possible Solutions*. Albertine Rift Technical Report Series No. 1: Wildlife Conservation Society, pp. 61–68.

———. 2002b. 'Primate Conservation and Local Communities—Ethical Issues and Debates', *American Anthropologist* 104(4): 1184–1194.

Hill, C.M. 2004. 'Farmers' Perspectives of Conflict at the Wildlife–Agriculture Boundary: Some Lessons Learned from African Subsistence Farmers', *Human Dimensions of Wildlife* 9(4): 279–286.

———. 2005. 'People, Crops, and Primates: A Conflict of Interests', in J.D. Paterson and J. Wallis (eds), *Commensalism and Conflict: The Human-Primate Interface*. Norman, OK: American Society of Primatologists, pp. 40–59.

———. 2015. 'Perspectives of "Conflict" at the Wildlife-Agriculture Boundary: 10 Years On', *Human Dimensions of Wildlife* 20(4): 296–301.

Hill, C.M., F.V. Osborn and A.J. Plumptre. 2002. 'Human-Wildlife Conflict: Identifying the Problem and Possible Solutions'. Albertine Rift Technical Report Series No. 1: Wildlife Conservation Society.

Hill, C.M., and G.E. Wallace. 2012. 'Crop Protection and Conflict Mitigation: Reducing the Costs of Living alongside Non-Human Primates', *Biodiversity and Conservation* 21(10): 2569–2587.

Hoare, R.E. 1995. 'Options for the Control of Elephants in Conflict with People', *Pachyderm* 19: 54–63.

Horrocks, J.A., and J Baulu. 1988. 'Effects of Trapping on the Vervet (*Cercopithecus aethiops sabaeus*) Population in Barbados', *American Journal of Primatology* 15: 223–233.

Hsiao, S.S., et al. 2013. 'Crop-Raiding Deterrents around Budongo Forest Reserve: An Evaluation through Farmer Actions and Perceptions', *Oryx* 47(4): 569–577.

Hulme, D., and M.W. Murphree (eds). 2001. *African Wildlife and Livelihoods: The Promise and Performance of Community Conservation*. Oxford: James Curry.

Hutton, J.M., and N. Leader-Williams. 2003. 'Sustainable Use and Incentive-driven Conservation: Realigning Human and Conservation Interests', *Oryx* 37(2): 215–226.

Hygnstrom, S.E., R.M. Timm and G.E. Larson (eds). 1994. *Prevention and Control of Wildlife Damage*. Washington, DC: Natural Resources and Rural Development Unit, US Department of Agriculture.

Infield, M. 1988. 'Attitudes of a Rural Community towards Conservation and a Local Conservation Area in Natal, South Africa', *Biological Conservation* 45(1): 21–46.

Kagiri, J. 2002. 'Human-Wildlife Conflicts in Kenya: A Conflict Resolution Concept', in C.M. Hill, F.V. Osborn and A.J. Plumptre (eds), *Human-Wildlife Conflict: Identifying the Problem and Possible Solutions*. Albertine Rift Technical Report Series No. 1: Wildlife Conservation Society, pp. 44–47.

Kapila, S., and F. Lyon. 1994. *Expedition Field Techniques: People Oriented Research*. London: Expedition Advisory Centre, Royal Geographical Society.

Kepe, T., B. Cousins and S. Turner. 2001. 'Resource Tenure and Power Relations in Community Wildlife: The Case of Mkambati Area, South Africa', *Society and Natural Resources* 14(10): 911–925.

Knickerbocker, T.J., and J. Waithaka. 2005. 'People and Elephants in the Shimba Hills, Kenya', in R. Woodroffe, S. Thirgood and A. Rabinowitz (eds), *People and Wildlife: Conflict or Coexistence?* Cambridge: Cambridge University Press, pp. 224–238.

Lane, S.J., and H. Higuchi. 1998. 'Efficacy of Common Protection Devices in Preventing Night-Time Damage of Cabbage Crops by Spot-Billed Ducks in Japan', *International Journal of Pest Management* 44(1): 29–34.

Lauridsen, M. 1999. 'Workers in a Forest: Understanding the Complexity of Incorporating Local People in Modern Management. A Case Study of the Nyabyeya Parish in Western Uganda', MSc thesis, University of Copenhagen.

Lee, P.C., and N.E.C. Priston. 2005. 'Human Attitudes to Primates: Perceptions of Pests, Conflict and Consequences for Primate Conservation', in J.D. Paterson and J. Wallis (eds), *Commensalism and Conflict: The Human-Primate Interface*. Norman, OK: American Society of Primatologists, pp. 1–23.

MacKenzie, C.A., and P. Ahabyona. 2012. 'Elephants in the Garden: Financial and Social Costs of Crop Raiding', *Ecological Economics* 75(1): 72–82.

Madden, F. 2004. 'Creating Coexistence between Humans and Wildlife: Global Perspectives on Local Efforts to Address Human-Wildlife Conflict', *Human Dimensions of Wildlife* 9(4): 247–257.

Maples, W.R., et al. 1976. 'Adaptations of Crop-Raiding Baboons in Kenya', *American Journal of Physical Anthropology* 45(2): 309–315.

Messmer, T.A. 2000. 'The Emergence of Human-Wildlife Conflict Management: Turning Challenges into Opportunities', *International Biodeterioration & Biodegradation* 45(3–4): 97–102.

Miller, K.K., and T.K. McGee. 2001. 'Toward Incorporating Human Dimensions Information into Wildlife Management Decision-Making', *Human Dimensions of Wildlife* 6(3): 205–221.

Morton-Williams, J. 1993. *Interviewer Approaches*. Aldershot: Dartmouth Publishing Company.

Mukherjee, A., and C.K. Borad. 2004. 'Integrated Approach towards Conservation of Gir National Park: The Last Refuge of Asiatic lions, India', *Biodiversity and Conservation* 13: 2165–2182.

Naughton-Treves, L. 1997. 'Farming the Forest Edge: Vulnerable Places and People around Kibale National Park, Uganda', *Geographical Review* 87(1): 27–46.

———. 1998. 'Predicting Patterns of Crop Damage by Wildlife around Kibale National Park, Uganda', *Conservation Biology* 12(1): 156–168.

———. 2001. 'Farmers, Wildlife, and the Forest Fringe', in W. Weber et al. (eds), *African Rain Forest Ecology and Conservation*. New Haven, CT: Yale University Press, pp. 588.

Newmark, W.D., et al. 1993. 'Conservation Attitudes of Local People Living Adjacent to Five Protected Areas in Tanzania', *Biological Conservation* 63(2): 177–183.

Nyhus, P.J., R. Tilson and Sumianto. 2000. 'Crop-Raiding Elephants and Conservation Implications at Way Kambas National Park, Sumatra, Indonesia', *Oryx* 34(4): 262–274.

O'Connell-Rodwell, C.E., et al. 2000. 'Living with the Modern Conservation Paradigm: Can Agricultural Communities Co-exist with Elephants? A Five-year Case Study in East Caprivi, Namibia', *Biological Conservation* 93(3): 381–391.

Osborn, F.V. 2002. 'Capsicum oleoresin as an Elephant Repellent: Field Trials in the Communal Lands of Zimbabwe', *Journal of Wildlife Management* 66(3): 674–677.

Osborn, F.V., and C.M. Hill. 2005. 'Techniques to Reduce Crop Loss: Human and Technical Dimensions in Africa', in R. Woodroffe, S. Thirgood and A. Rabinowitz (eds), *People and Wildlife: Conflict or Coexistence?* Cambridge: Cambridge University Press, pp. 72–85.

Osborn, F.V., and G.E. Parker. 2002a. 'Living with Elephants II: A Manual for Implementing an Integrated Programme to Reduce Crop Loss to Elephants and to Improve Livelihood Security of Small-Scale Farmers', *Mid Zambezi Elephant Project*, pp. 21.

———. 2002b. 'An Integrated Approach toward Problem Animal Management', in C.M. Hill, F.V. Osborn and A.J. Plumptre (eds), *Human-Wildlife Conflict: Identifying the Problem and Possible Solutions*. Albertine Rift Technical Report Series No. 1: Wildlife Conservation Society, pp. 121–127.

———. 2003. 'Towards an Integrated Approach for Reducing the Conflict between Elephants and People: A Review of Current Research', *Oryx* 37(1): 80–84.

Padulosi, S., P. Eyzaquirre and T. Hodgkin. 1999. 'Challenges and Strategies in Promoting Conservation and Use of Neglected and Underutilized Crop Species', in J. Janick (ed.), *Perspectives on New Crops and New Uses*. Alexandria, VA: ASHS Press, pp. 140–149.

Paterson, J.D., and J. Wallis (eds). 2005. *Commensalism and Conflict: The Human-Primate Interface*. Norman, OK: American Society of Primatologists.

Peterson, M.N., et al. 2010. 'Rearticulating the Myth of Human–Wildlife Conflict', *Conservation Letters* 3(2): 74–82.

Pirta, R.S., M. Gadgil and A.V. Kharshikar. 1997. 'Management of the Rhesus Monkey *Macaca mulatta* and Hanuman Langur *Presbytis entellus* in Himachal Pradesh, India', *Biological Conservation* 79(1): 97–106.

Plumptre, A.J. 2002. 'Crop Raiding around the Parc National des Volcans, Rwanda: Farmer's Attitudes and Possible Links with Poaching', in C.M. Hill, F.V. Osborn and A.J. Plumptre (eds), *Human-Wildlife Conflict: Identifying the Problem and Possible Solutions*. Albertine Rift Technical Report Series No. 1: Wildlife Conservation Society, pp. 79–88.

Plumptre, A.J., et el. 2004. 'The Socio-economic Status of People Living Near Protected Areas in the Central Albertine Rift.' Albertine Rift Technical Report Series No. 4: Wildlife Conservation Society, pp. 132.

Plumptre, A.J., and V. Reynolds. 1994. 'The Effect of Selective Logging on the Primate Populations in the Budongo Forest Reserve, Uganda', *Journal of Applied Ecology* 31(4): 631–641.

Priston, N.E.C., and M.R. McLennan. 2013. 'Managing Humans, Managing Macaques: Human-Macaque Conflict in Asia and Africa', in S. Radhakrishna, M. Huffman and A. Sinha (eds), *The Macaque Connection: Cooperation and Conflict between Humans and Macaques*. New York: Springer, pp. 225–250.

Randall, S., and T. Koppenhaver. 2004. 'Qualitative Data in Demography: The Sound of Silence and Other Problems', *Demographic Research* 11(3): 57–94.

Rao, M., and P.J.K. McGowan. 2002. 'Wild-Meat Use, Food Security, Livelihoods, and Conservation', *Conservation Biology* 16(3): 580–583.

Redpath, S.M., et al. 2013. 'Understanding and Managing Conservation Conflicts', *Trends in Ecology and Evolution* 28(2): 100–109.

Riley, S.J., et al. 2003. 'Adaptive Impact Management: An Integrative Approach to Wildlife Management', *Human Dimensions of Wildlife* 8(2): 81–95.

Rode, K.D., et al. 2006. 'Nutritional Ecology of Elephants in Kibale National Park, Uganda, and its Relationship with Crop-Raiding Behaviour', *Journal of Tropical Ecology* 22: 441–449.

Rollins, K., and H.C. Briggs. 1996. 'Moral Hazard, Externalities, and Compensation for Crop Damages from Wildlife', *Journal of Environmental Economics and Management* 31(3): 368–386.

Rollinson, D. 2005. *Organisational Behaviour and Analysis: An Integrated Approach*, 3rd ed. Harlow: Prentice Hall.

Sitati, N.W., and M.J. Walpole. 2006. 'Assessing Farm-Based Measures for Mitigating Human-Elephant Conflict in Transmara District, Kenya', *Oryx* 40(3): 279–286.

Sitati, N.W., M.J. Walpole and N. Leader-Williams. 2005. 'Factors Affecting Susceptibility of Farms to Crop Raiding by African Elephants: Using a Predictive Model to Mitigate Conflict', *Journal of Applied Ecology* 42: 1175–1182.

Soto, B., S.M. Munthali and C. Breen. 2001. 'Perceptions of the Forestry and Wildlife Policy by the Local Communities Living in the Maputo Elephant Reserve, Mozambique', *Biodiversity and Conservation* 10: 1723–1738.

Stewart, G.B., C.F. Coles and A.S. Pullin. 2005. 'Applying Evidence-Based Practice in Conservation Management: Lessons from the First Systematic Review and Dissemination Projects', *Biological Conservation* 126(2): 270–278.

Strum, S.C. 1986. 'Activist Conservation: The Human Factor in Primate Conservation in Source Countries', in J.G. Else and P.C. Lee (eds), *Primate Ecology and Conservation*. Cambridge: Cambridge University Press, pp. 367–382.

———. 1994. 'Prospects for Management of Primate Pests', *Revue d'Ecologie (La Terre et La Vie)* 49(3): 295–306.

Thirgood, S., R. Woodroffe and A. Rabinowitz. 2005. 'The Impact of Human-Wildlife Conflict on Human Lives and Livelihoods', in R. Woodroffe, S. Thirgood and A. Rabinowitz (eds), *People and Wildlife: Conflict or Coexistence?* Cambridge: Cambridge University Press, pp. 13–26.

Toutain, B., M.-N. De Visscher and D. Dulieu. 2004. 'Pastoralism and Protected Areas: Lessons Learned from Western Africa', *Human Dimensions of Wildlife* 9(4): 287–295.

Tungittiplakorn, W., and P. Dearden. 2002. 'Biodiversity Conservation and Cash Crop Development in Northern Thailand', *Biodiversity and Conservation* 11(11): 2007–2025.

Tweheyo, M., C.M. Hill and J. Obua. 2005. 'Patterns of Crop Raiding by Primates around the Budongo Forest Reserve, Uganda', *Wildlife Biology* 11(3): 237–247.

Van Huis, A., and F. Meerman. 1997. 'Can we Make IPM Work for Resource-Poor Farmers in Sub-Saharan Africa?' *International Journal of Pest Management* 43(4): 313–320.

Wallace, G.E. 2010. 'Monkeys in Maize: Primate Crop-Raiding Behaviour and Developing On-farm Techniques to Mitigate Human-Wildlife Conflict', PhD thesis, Oxford Brookes University, Oxford, pp. 528.

Wallace, G.E., and C.M. Hill. 2012. 'Crop Damage by Primates: Quantifying the Key Parameters of Crop-Raiding Events', *PLoS ONE* 7(10): e46636.

Warren, Y. 2003. 'Olive Baboons (Papio cynocephalus anubis): Behaviour, Ecology and Human Conflict in Gashaka Gumti National Park, Nigeria', PhD thesis, University of Surrey, Roehampton, pp. 308.

Warren, Y., B. Buba and C. Ross. 2007. 'Patterns of Crop-Raiding by Wild and Domestic Animals near Gashaka Gumti National Park, Nigeria', *International Journal of Pest Management* 53(3): 207–216.

Webber, A.D. 2006. 'Primate Crop Raiding in Uganda: Actual and Perceived Risks around Budongo Forest Reserve', PhD thesis, Oxford Brookes University, Oxford.

Webber, A.D., and C.M. Hill. 2014. 'Using Participatory Risk Mapping (PRM) to Identify and Understand People's Perceptions of Crop Loss to Animals in Uganda', *PLoS ONE* 9(7): e102912.

Webber, A.D., C.M. Hill and V. Reynolds. 2007. 'Assessing the Failure of a Community-Based Human-Wildlife Conflict Mitigation Project in Budongo Forest Reserve, Uganda', *Oryx* 41(2): 177–184.

Woodroffe, R., S. Thirgood and A. Rabinowitz (eds). 2005. *People and Wildlife: Conflict or Coexistence?* Cambridge: Cambridge University Press.

WWF. 2005. *Human Wildlife Conflict Manual*. Harare: World Wide Fund for Nature, Southern Africa Regional Programme.

Young, J., et al. 2005. 'Towards Sustainable Land Use: Identifying and Managing the Conflicts between Human Activities and Biodiversity Conservation in Europe', *Biodiversity and Conservation* 14(7): 1641–1661.

10
Using Geographic Information Systems at Sites of Negative Human-Wildlife Interactions
Current Applications and Future Developments

Amanda D. Webber, Stewart Thompson, Neil Bailey and Nancy E.C. Priston

Negative interactions between humans and wild animals are a major challenge to biodiversity conservation globally (Hill, Osborn and Plumptre 2002; Woodroffe, Thirgood and Rabinowitz 2005; Karanth et al. 2012). Competition for space and resources is intensifying due to ever-expanding human populations and increasing conversion of natural habitats to alternative land uses (Thirgood, Woodroffe and Rabinowitz 2005; Hoffman and O'Riain 2012). Some wildlife species will adapt and potentially benefit from being in human-dominated environments (e.g. Gautier and Biquand 1994; Le Lay, Clergeau and Hubert-Moy 2001; Kretser, Curtis and Knuth 2009). Indeed, anthropogenic landscapes are vital to the survival of some open habitat species (Wright, Lake and Dolman 2012). However, taxa that cannot adjust to more agricultural and, in some cases, urbanized environments face local extinction. Conversely, for sustainable coexistence to be achievable, people may need to increase their tolerance towards wildlife. To avoid a hostile interface with local human populations we must understand the ecological and geographical drivers of human-wildlife interactions (HWI) and, if we are to target management strategies effectively and efficiently, we must be able to locate mitigation measures precisely (Malo, Suárez and Diez 2004).

GIS (Geographic Information Systems) technology is a flexible and systematic method of integrating large quantities of data through a spatial reference system that provides tools to explore the data (Haslett 1990; Michelmore 1994; Randy-Gimblett 2002). A GIS is often used to present and/or analyse data that have been collected with GPS (Global Positioning System)–enabled devices or remotely sensed data from satellites (Hughes 2003; Hebblewhite and Haydon 2010; Tomkiewicz et al. 2010). These spatial data enable researchers to identify likely sites of negative interaction and mitigation in addition to model and predict scenarios without disturbing vulnerable landscapes (Randy-Gimblett 2002). The complexity of these interactions, which often involve multiple spatial and temporal factors, lend themselves to mapping using GIS, and we are currently seeing a proliferation of research in this area.

In this chapter we examine how GIS has been used at sites of HWI, and use case studies to explore the efficacy of this approach. In conclusion, we consider limitations and new developments of GIS technology in this vital area of conservation.

Using GIS at Sites of Human-Wildlife Interaction

The use of GIS in conservation has become more prevalent in recent years (see Le Lay, Clergeau and Hubert-Moy 2001; Sitati et al. 2003; Enari and Suzuki 2010; Tomkiewicz et al. 2010). While much previous work examining HWI has focused on elephants and carnivores (see Smith and Kasiki 2000; Sitati et al. 2003; Treves et al. 2004), mapping of HWI in human-dominated landscapes has also taken place (e.g. Le Lay, Clergeau and Hubert-Moy 2001; Malo, Suárez and Diez 2004). While GIS is often used to predict future risk or vulnerability, in this chapter we will begin by outlining how this technology has been used to visualize an existing situation.

Visualizing Human-Wildlife Interaction

Overlaying maps of species habitat or behaviour with those of protected areas or human activity is one of the simplest and most frequent applications of GIS at sites of HWI. This process has enabled users to identify areas at high risk from increased competition between people and animals (see Fox, Yonzon and Podger 1996). For example, Alan Rabinowitz and Katherine Zeller (2010) identified landscape features vital for jaguar (*Panthera onca*) movement in the wild and asked experts to assign cost values to specific attributes that included human population density and distance from roads and settlements. By merging the feature layers in a GIS, a map was created that could identify potential movement corridors and barriers; a vital tool for the conservation planning of this species. However, it is important to note that co-occurrence does not automatically result in 'conflict' (Waldron et al. 2013); rather this technique highlights *potential* locations of negative human-wildlife interaction.

In addition to identifying where HWI may occur, GIS mapping has been used to evaluate the success of specific mitigation strategies (that have spatial significance). For example, Robert Smith and Samuel Kasiki (2000) assessed the efficacy of an electric fence as a method of reducing negative interactions with elephants around Tsavo, Kenya. GIS maps and analysis revealed that the fence had no significant impact on the absolute number of incidents nor did it alter the distribution of events. In this example, further intervention was required to (1) reduce negative interactions by extending the length of the fence so animals could not easily circumnavigate the barrier and (2) encourage farmers to avoid planting crops close to the fence boundary. It is important to note that the latter strategy may be difficult to accomplish if land is scarce (e.g. Mwavu and Witkowski 2008).

GIS has also been used to integrate biological and social science data to examine the human dimensions of wildlife management. Overlaying biological landscape maps with those of perceptions of HWI has helped practitioners to understand whether interaction between people and wildlife was likely to be negative (Kretser, Curtis and Knuth 2009). For example, Brett Anthony Bryan et al. (2010) examined the correlation between ecological and social values (i.e. natural capital assets and ecosystems services) across natural areas in southern Australia and found only a small proportion of the landscape had high scores for both factors. This has significance for conservation planning as areas with a divergence between values may require increased engagement with local people and/or incentive programmes. Mapping biological and social data could also identify where mitigation strategies would be both effective and acceptable for local people. While creating visual representations of human-wildlife interaction clearly has value, GIS has also been used to explore data and manipulate outcomes using spatial modelling.

Predicting Human-Wildlife Interaction

Natural hazards, such as crop damage by wildlife, are not random (Treves et al. 2011); consequently, spatial modelling, with its ability to identify future areas of resource overlap and/or negative interaction with people, is an important development in the evolution of effective mitigation strategies.

Predictive spatial models or risk maps have been developed based on factors such as proximity to reserves and/or habitat refuges, roads, types of protection strategy employed and human activity (e.g. Sitati et al. 2003; Treves et al. 2004; Wilson et al. 2006; Ahmadi et al. 2013; Behdarvand et al. 2014). For example, GIS has been used to model the risk to both agricultural and property damage by a recovering population of Japanese macaques (*Macaca fuscata*); maps revealed that existing mitigation may not be enough to address the issue of expanding populations and precautionary measures may be required to reduce future negative interactions with people (Enari and Suzuki 2010). Seth Wilson et al. (2006) used GIS to evaluate the impact of electric fencing as a mitigation strategy to prevent beehive damage by grizzly bears (*Ursus arctos*). Surprisingly they found that fencing was associated with an increased likelihood of damage. Further research is required to understand if bears are 'testing' fences or results are due to associations with other attractants nearby (Wilson et al. 2006). GIS has even been used to try and ascertain factors that influence people's perceptions of wildlife, e.g. black bears (*Ursus americanus*) (Kretser, Curtis and Knuth 2009), and to map perceived interactions and compensation distribution (Karanth et al. 2012). Krithi Karanth et al. (2012) found that no single intervention was effective in lowering perceived livestock or crop loss around a protected area in central India, suggesting that mitigation in this case may only be effective if targeted to individual households rather than generalized at the community level. Predictive models can also show where human behaviour needs to be modified to prevent or minimize negative human-wildlife interaction; for example, grazing cattle in areas that are identified as being of high risk of livestock predation should be avoided (Wilson et al. 2006, Treves et al. 2011, Behdarvand et al. 2014). These insights are vital for conservation planning where limited funds and resources necessitate the prioritization of investment (Kahler, Roloff and Gore 2012). However, predictive models should not be relied upon without ground truthing to validate assumptions. Factors may be missing from the analysis and, while models can show correlations between factors, they cannot present causal relationships (Treves et al. 2011; Ahmadi et al. 2013).

Stakeholder Engagement and Participatory GIS

Presenting complex HWI data in a way that can be easily interpreted by all stakeholders is very important for successful mitigation (Smith and Kasiki 2000). Geographic data have more visual power and are more easily understood by stakeholders if visually presented as a map (Lewis 1995; Cinderby 1999; Elwood 2006; Hebblewhite and Haydon 2010), which is advantageous in areas where illiteracy is common. Using GIS images in this way can both engage and build trust between stakeholders (Le Lay, Clergeau and Hubert-Moy 2001; Brown and Reed 2009; Brown and Weber 2011) as many negative HWIs are exacerbated by tensions in the relationships between wildlife authorities and local people (e.g. Tumusiime and Svarstad 2011). Such images have the potential to present, an often emotive subject, in objective and authoritative terms; as a local leader stated in a study from Zambia, 'our maps were like one's heart, because they spoke with truth and people believed in what they said' (Ndonyo in Lewis 1995: 867).

Participatory GIS (PGIS) uses GIS technology to 'produce local knowledge with the goal of including and empowering marginalised populations' (Brown and Reed 2009: 166–167); historically, formal maps have often been facilitated and/or owned by outsiders (Chambers 2006). Specific spatial skills and training are not necessarily required, as residents can annotate paper base maps or be supplied with mobile phones, contributing valuable local information (Lewis 1995). Online and mail surveys have also been used (Brown and Weber 2011); for example, Cody Cox et al. (2014) asked participants to affix stickers to a map to show their opinions on conservation priority areas (that were then overlaid with habitat maps). However, these data will still need to be digitized, which is labour intensive, expensive and a top-down approach (Cinderby 1999; Brown and Reed 2009), or appropriate data architecture will be required to semi-automate the process. As Dale Lewis describes, 'it is remote sensing with humans taking the place of satellites, and gathering information about the location of individual villages, their histories and demographics, and potential threats to surrounding resources' (1995: 865). One criticism is that there always has to be a geo-referenced map underneath and this may restrict communities; they may have a different concept of space and their landscape (Cinderby 1999; Carver 2003).

PGIS is also community specific and can be difficult to transfer to other places without a thorough understanding of how the method is 'situationally and culturally influenced' (Sieber 2006: 495) – for example, the local context, stakeholders involved and possible legal issues with data. It is important to be aware that the benefits of GIS, 'such as improved visualization or data accuracy, can induce further injustice' (Sieber 2006: 495) for already marginalized, local communities; e.g. in Tanzania, mapping was regarded with suspicion as it seemed to support a historical challenge to a pastoral way of life (Hodgson and Schroeder 2002). It needs to be used as a process and not a goal or it could increase the potential for social conflict, e.g. at sites with disputed or poorly defined land ownership (Cinderby 1999).

A development on this approach is Participatory 3D Modelling (P3DM), where a 3D model is created from topographic information. Local people create the map and all legend information, and if a scaled grid is added data can be input into a GIS (Rambaldi 2010). It makes content 'portable and shareable' and its vertical features make it a more effective stimulus for discussion than a 2D sketch map (Rambaldi 2010: 5). This approach is particularly useful at sites with numerous conflicting interests (e.g. negative HWI), as it is a tangible format that can engage all stakeholders in meaningful discussion.

To elucidate the potential and range of applications for GIS at sites of HWI, we present three case studies from Uganda, the United States and Kenya.

Understanding Patterns of Crop Damage by Primates around Budongo Forest Reserve, Uganda

In Uganda, GIS mapping proved to be an effective way of visualizing crop damage by wild vertebrates around Budongo Forest Reserve (BFR). There is a history of negative interactions between people and wild animals in this area, particularly primates, e.g. olive baboons (*Papio anubis*), chimpanzees (*Pan troglodytes schweinfurthii*) and red-tailed monkeys (*Cercopithecus ascanius*) (see Hill 1997; Tweheyo, Hill and Obua 2005; Webber and Hill 2014). This is driven by the conversion of many forest fragments to agriculture, restrictions on lethal crop-protection strategies and, following the loss of employment opportunities, a reliance on farming as a sole source of income for the majority of local people (Hill 1997; Webber 2006; Mwavu and Witkowski 2008). Negative HWIs are seldom evenly distributed (Woodroffe, Thirgood and Rabinowitz 2005) and, at this site, only a small percentage of farmers experienced crop damage on a regular basis (Webber 2006).

GIS maps clearly indicated that areas vulnerable to damage events by wild species, particularly primates, were disproportionally represented at the forest edge. This was unsurprising, as many large vertebrates live, sleep and forage in forest habitats. However, the maps also highlighted subtle details that may not have been obvious from the data points alone and yet are significant for the mitigation of crop loss. Primate damage events were close to the forest edge but animals did not seem to be deterred by crossing a non-forested zone between BFR and cultivated land. Ground truthing revealed this area to be bushy vegetation (Webber 2005, pers. obs.). Once in the farms however, damage was generally close to the farm boundary. It suggests that a buffer zone of non-palatable vegetation was not enough to protect farms from incursions by wild primate species. This example demonstrates how GIS maps of vulnerability can be used to highlight where interaction (in this case crop damage) may occur and where mitigation strategies, e.g. guard huts or early warning devices, would be best placed (Sitati et al. 2003; Webber 2006). It should be noted that the crop-foraging behaviour of different animals at this site displayed specific spatial patterning. For example, bushpig (*Potamochoerus* spp.) damage often occurred at a greater distance from the forest edge than that of other wildlife and incursions went deeper into farm land. This may be facilitated by nocturnal foraging behaviour and the lack of guarding by local people at this time (Webber 2006). It demonstrates how the type and location of mitigation strategies at this site may need to be modified to the 'problem' species in question.

Developing a Participatory Approach to Human–Grizzly Bear Interactions in Montana

Repeated negative interactions between large carnivores and humans usually lead to the permanent removal of the offending animal by relocation or killing (Treves and Karanth 2003; Wilson, Neudecker and Jonkel 2014). In the Blackfoot River Watershed, Montana, a growing population of grizzly bears (*Ursus arctos*), and associated range expansion, has led to an increase in negative interactions between people and these large carnivores, resulting in one human fatality (Wilson, Neudecker and Jonkel 2014). In 2001, a local NGO (Blackfoot Challenge) and the Montana Fish, Wildlife and Parks (MFWP) met with local people to try and understand the impact of these changes on the predominately ranching community.

GIS was used to map grizzly bear–human interactions (from data provided by MFWP) to identify 'conflict hotspots' (Wilson et al. 2005). It was also used to model the probability of negative interactions occurring depending on the presence of predictor variables, i.e. riparian vegetation and bear 'attractants' such as beehives and 'boneyards' (disposal areas for carcasses of calves that die during the calving season) (Wilson et al. 2005, 2006). The GIS component was part of a holistic strategy to address the issue that included surveys to assess perceptions towards bears; these showed there were multiple perspectives and the community was divided as to whether Blackfoot would be a better place to live without grizzly bears on private land (Wilson, Neudecker and Jonkel 2014). While the biological maps were an essential mitigation tool, perhaps the most important role of the GIS in this project was during the subsequent participatory spatial mapping workshops. Stakeholders were asked to use the maps to discuss interactions, outline land-use practices and plot the location of known bear attractants. As Seth Wilson, Gregory Neudecker and James Jonkel (2014) state, using GIS maps in this way gave a focus to move beyond the polarized perceptions of bear interactions and to look to solutions. It was an effective way to engage with the community and for all stakeholders to explore and understand the motivations of others. It was also a valuable tool for developing 'trust' by sharing and 'mutual learning'; 'mapping reverses the traditional flow of information in practical problem solving' from a top down to a bottom up approach (Wilson, Neudecker and Jonkel 2014: 188).

The participatory process took time, patience and funding but it did lead to community-level goals that were relevant, appropriate and achievable by stakeholders. These goals were framed by stakeholder identity as rural landowners, and focussed on modifying human behaviours; e.g. key areas to address were to prevent interactions (through electric fences, waste management and livestock carcass removal), protect human safety and guard against livestock loss (through a neighbour-neighbour communication network) (Wilson, Neudecker and Jonkel 2014). The efficacy of this participatory approach (including GIS) to mitigation has been demonstrated by a 96 per cent reduction in negative human-bear interactions from 2003–2010 (Wilson, Neudecker and Jonkel 2014). While bear management remains a priority for stakeholders in this area, Blackfoot Challenge is looking to build on this success by examining negative interactions between ranchers and wolves (Wilson, Neudecker and Jonkel 2014).

Geo-Fencing Elephants in Kenya

In Kenya, GIS has been used to facilitate an early warning system to enhance coexistence between elephants (*Loxodonta africana*) and people. Here, as in other parts of their range, elephants have conflicting status; from 'landscape gardeners', symbols of wilderness and sources of income, to crop foragers and destroyers of property (Douglas-Hamilton, Krink and Vollrath 2005). Established approaches for dealing with these negative interactions include shooting 'problem animals' and the erection of physical barriers, i.e. electric fences.

From the mid-1990s, a small number of elephants in Kenya were fitted with radio collars to monitor their movements. These collars allow for high-resolution GPS tracking with a regular fix on location (usually one or three hours) (Douglas-Hamilton, Krink and Vollrath 2005). Studies revealed that elephants spend a large proportion of their time outside protected areas and many movement corridors are in these human-dominated zones; elephants often forage at night and move faster in these areas to minimize the risk of interacting with people (Douglas-Hamilton, Krink and Vollrath 2005; Graham et al. 2009). In more recent years, collars have been fitted with a GSM (Global System for Mobile Communication) modem to enable two-way communication and remote programming for the GPS receivers (Graham et al. 2009).

In 2007 a radio-collared bull elephant embarked on an extensive period of crop-foraging and damage to electric fencing. To avoid culling the individual, a real time virtual fence (RTVF), or 'geo-fence', was introduced (Wall et al. 2014). As Jake Wall et al. (2014) describe, managers were alerted via SMS, with GPS coordinates, whenever the bull elephant crossed key spatial points in the GIS, in this case the location of the electrified perimeter fence. Patrol teams responded to these incursions and through negative reinforcement the individual was deterred from entering the agricultural zone; since 2008 this elephant has not been recorded foraging on crops (Save the Elephants 2016).

Geo-fences are, in contrast with physical barriers, specific to the individuals concerned. They are also flexible because modification of the boundaries can occur without any physical interaction with the landscape, and facilitate a rapid response (Jachowski, Slotow and Millspaugh 2014). This use of GIS technology has successfully kept radio-collared elephants from sites of negative HWI and is being introduced across other sites in Africa (Save the Elephants 2016). In addition to being used to deter individuals, it can also be used to alert and direct anti-poaching patrols once elephants pass through geographic boundaries or if a collared animal becomes immobile. This approach has also been useful as a public engagement tool; in January 2014, four elephants had their GPS data uploaded to twitter (Space for Giants 2014).

Limitations of Using GIS in HWI Scenarios

The use of GIS is becoming more important at sites of HWI, yet it is not without problems that may prove challenging to both academic researchers and conservation practitioners. Collecting the appropriate spatial data can be difficult, as GPS devices are limited by battery power and cannot always give an accurate location quickly (Tomkiewicz et al. 2010). For example, there can still be problems getting a timely fix on location if the path to satellites from the receiver is obscured by dense tree cover (Hughes 2003; Frair et al. 2010). Additionally, to obtain meaningful results a large volume of data is required (Koenig 1999). Collecting measurements in the field (ground truthing) is time consuming and, as landscapes are constantly evolving, it may need to be repeated a number of times depending on the factor under examination (Haslett 1990; Hughes 2003; Rabinowitz and Zeller 2010). Ironically, time saving is a major advantage of using GPS technologies in other scenarios, e.g. radio telemetry (Hebblewhite and Haydon 2010). Ideally, any site will need to be surveyed over a long period in order to gain the volume of data required, but for conservation-based projects this is not always practical or economically viable.

As surveying relies on humans or remotely sensed data, e.g. satellite imagery, there is also potential for inputting errors. This is made more likely as spatial data are often obtained at different scales (as they come from a number of different sources) and may need to be converted/manipulated before they can be used (Smith and Kasiki 2000; Anderson and Gaston 2013). This is another time-consuming, and therefore expensive, process (Haslett 1990; Smith and Kasiki 2010; Hughes 2003). Furthermore, due to the necessity of accessing software and reliable internet connections, data entry and digitizing are frequently conducted away from the field site so inaccurate or spurious points are not identified until it is too late.

HWIs often occur in locations that are environmentally and socially fragile, with marginalized, economically poor human communities competing for resources with protected wildlife populations (McLennan and Hill 2012). The hardware and software required to create GIS are expensive and to conduct more advanced analysis, individuals require specific modelling and statistical skills (Carver 2003; Sieber 2006; Hebblewhite and Haydon 2010; Rambaldi 2010). There is growing expertise in developing areas (e.g. Africa) and free software is more readily available (see GRASS GIS 2016), but projects are hampered by unreliable internet access and the need for advanced experience of spatial analysis (Lewis 1995; Carver 2003; Elwood

2006; Swetnam and Reyers 2011). To address this potential limitation the GIS component of a given project is often either never taken beyond simplistic levels or else is managed by external, international agencies/funders. Even if it is based in the local community with a support network available it is 'likely to be informal and fragile and fail to ensure long-term sustainability of the GIS' (Sieber 2006: 499).

Using GIS, and particularly modelling spatial data, can be complex; selecting appropriate methods and tools can be problematic and analysis and interpretation of the data is dependent on the user's ability and the quality of the data as opposed to the GIS technology itself (Michelmore 1994). For robust statistical analysis, grid cells need to be independent of one another or spatial autocorrelation (SA) can occur, potentially inflating the likelihood of type 1 statistical errors (Koenig 1999; Smith and Kasiki 2000; Sitati et al. 2003; Frair et al. 2010). This requires a compromise; resolution has to be coarse enough to allow for independence, but not too coarse or accuracy may be lost along with spatial patterns. One option is to select GPS points randomly to use in analysis but this does not always eradicate the problem (Bryan et al. 2010; Hoffman and O'Riain 2012). SA, therefore, can be a particular problem for results at a fine level of analysis, which, unfortunately, is exactly the level at which data are required to demonstrate the link between resource and behaviour (Smith and Kasiki 2000; Sitati et al. 2003; Hebblewhite and Haydon 2010). As many sites of HWI cover localized issues, especially those examining crop damage by wildlife with small ranges (e.g. some primates), this will require specific consideration. Mark Boyce et al. (2010) suggest that spatial autocorrelation should be considered an important element of the data, worthy of greater attention by researchers. Noah Sitati et al. (2003: 675) argue in their analysis of spatial patterns of human-elephant interactions at different resolutions that 'a compromise in resolution on statistical grounds does not affect the identification of underlying relationships and may improve clarity by reducing noise'. However, researchers may need to accept that there are potential limitations to mapping in smaller communities (Enari and Suzuki 2010). One also needs to consider issues raised by the comparison of spatial data across sites due to differences in the resolution of data (Michelmore 1994; Sitati et al. 2003; Cox et al. 2014).

Future Developments

We are witnessing technological developments that could make the collection of GIS data at sites of HWI more effective in the future. Smartphones and tablet devices are increasingly being used in developing countries as an alternative for personal computers (Marshall 2007); many have built-in GPS, and thus the capability to gather spatial data, and can now be purchased for less than £100. Devices such as these are increasingly used to enable indigenous people to gather data; for example, in Kenya, the Samburu warriors of the Ewaso Lions Project use mobile technology to report on wildlife presence and potential negative interactions in exchange for education classes (WildKnowledge 2011). Smartphones also have the ability to gather data offline and then upload from fixed Wi-Fi points, which reduces problems with unreliable internet access. This clearly has great potential, but barriers to producing advanced GIS analysis may be more difficult to resolve.

Unmanned Aerial Vehicles (UAVs), also known as Remotely Piloted Aircraft Systems (RPAS) or drones, are being increasingly discussed as potential tools for the collection of spatial data at sites of conservation interest (Schiffman 2014; Mulero-Pázmány et al. 2014; see also www.conservationdrones.org). Lightweight UAVs are small, capable of flying high enough to avoid disturbing wildlife (optimum 100–180 m) and do not need special sites to launch or land (Tenenbaum 2012/13; Anderson and Gaston 2013; Mulero-Pázmány et al. 2014). They are readily available, affordable and can be operated easily with little training (Tenenbaum 2012/13; Anderson and Gaston 2013; Schiffman 2014). As Karen Anderson and Kevin Gaston (2013) discuss, one of the biggest advantages, in the context of this research, is that UAVs can provide spatial data at the finer scales required for ecological research. Furthermore, there are currently few regulations in place for this technology beyond 'line of sight' flying and a need to remain in unpopulated areas (Anderson and Gaston 2013; Mulero-Pázmány et al. 2014). However, this may change in the future as (1) the number of amateur pilots rises and the potential risk to commercial airlines increases (Pigott 2014) and (2) governments become more concerned with risks to their local security (Cress and Zommers 2014).

With regard to sites of negative HWI, UAVs could have a number of applications. For example, they could be used to map migration corridors to ensure wildlife and people can coexist in areas of potential interaction (Schiffman 2014). They may also prove useful for monitoring mitigation strategies, i.e. electric fences; Margarita Mulero-Pázmány et al. (2014) found that images of a fence were acceptable at morning and midday surveillance test flights, so it would be possible to monitor damage from wildlife incursions remotely. If flown purposefully low, UAVs could also be used as a deterrent to keep wildlife away from human-dominated landscapes; elephants are known to avoid bees (King, Douglas-Hamilton and Vollrath 2011) and the sound of low-flying drones seems to have the same effect (Schiffman 2014). UAVs also provide real-time GPS coordinates so could ensure a timely response at a likely negative HWI (Tenenbaum 2012/13). This could have potential for conservation sites that require enforcement; e.g. UAVs could be used to monitor protected areas at risk from poaching and deter illegal activities (Tenenbaum 2012/13; Cress and Zommers 2014; Mulero-Pázmány et al. 2014; Schiffman 2014). However, the ethics of remotely monitoring human space use in HWI scenarios needs to be considered very carefully. There are concerns that UAVs will be viewed as 'sinister technologies of surveillance or be associated with warfare' (Humle et al. 2014: 1351). As Tatyana Humle et al. (2014) state, this type of technology might be seen as a return to 'fortress conservation', so may weaken the fragile trust that often exists between wildlife authorities, conservation organizations and local people around protected areas. While the ecological potential of UAVs in monitoring wildlife populations is clearly important, there needs to be further research into the acceptance of this technology by local people who live close to wildlife and by those who support conservation sites, e.g. tourists (Mulero-Pázmány et al. 2014).

Conclusion

Clearly, using GIS has advantages and disadvantages at sites of HWI. Identifying areas at high risk of negative interaction can both prevent problems developing through targeting of management strategies, and gain the support of local people for environmental initiatives (Treves et al. 2004). It also has the 'potential to engage and motivate decision-makers in unique ways' (Swetnam and Reyers 2011: 413). This is the 'holy grail' for those engaged in wildlife conservation and attempting to protect the assets of local human and wildlife populations. However, any benefits have to be balanced with potential economic and labour costs in addition to the need for specific technical skills. Nevertheless, as access to cheap mobile technology and online GIS continues to improve (Swetnam and Reyers 2011), spatial data gathering will become more affordable. This is encouraging, especially for a more widespread adoption of PGIS among local communities. The need for experts to analyse/manipulate advanced GIS data or train local people is a limitation that is presently difficult to resolve but perhaps that should not be a barrier for its use at sites of negative HWI. While complex modelling is of interest to the academic research community, spatial representation on a map is simple to produce and may be sufficient for community-based identification of risk and potential locations of mitigation strategies.

Another way of improving the potential of GIS projects at sites of HWI is by encouraging greater collaboration between all stakeholders (Wilson et al. 2006). As Karel Hughes outlines, there is a 'need to dissolve the boundaries that isolate scientists within their different disciplines' (2003: 69). From a research perspective, HWI (and conservation biology more generally) necessitates a more integrated approach. Larger, interdisciplinary teams working across several sites could pool expertise and also examine 'conflict' scenarios from a range of different perspectives, including spatial aspects. Projects that encourage researchers to share spatial information have demonstrated the potential efficacy of this method; e.g. the African Elephant Database is a GIS where international researchers can add and use spatial information on elephants and human-elephant conflict (Michelmore 1994; Smith and Kasiki 2000; African Elephant Specialist Group 2016). It would be useful to see this developed for other HWI issues or species that commonly forage on agricultural crops, e.g. baboons. While each site will have specific ecological, social and economic features, it is important to identify if there are wider patterns of communality across sites/species. This will be significant for our understanding of effective mitigation measures; tools or strategies trialled in one situation may be successful elsewhere.

John Haslett (1990) mused that applied ecology could be the area of research that might benefit significantly from GIS technology. We are only just beginning to understand its long-term contribution to the identification and mitigation of negative HWI, but with improved internet access, cheaper hardware/software and the continued development of mobile technology, it is hoped that its potential to provide a holistic view of risk through the integration of environmental and social factors will be realized.

Acknowledgements

For the Uganda fieldwork, a research permit was granted through the Uganda National Council for Science and Technology and the Uganda Wildlife Authority. The project was supported by an Oxford Brookes University Scholarship, a Wildlife Conservation Society Research Fellowship and grants from the Parkes Foundation, Wenner Gren Foundation, Primate Conservation Inc., a British Airways / Royal Geographical Society (with IBG) Travel Bursary and discounted equipment from Silva Ltd. We were grateful for the help and support of Ruthlen Atugonza, Jackson Okuti, Mawa Diedonne, Geoffrey Okethuwengu and the people of Masindi District. Thanks also to Dr Nadine Laporte of Woods Hole Research Centre and United Nations FAO Africover Project for providing maps and assistance with GIS software. We also thank Drs Matt McLennan and Kim Hockings for their comments on earlier drafts of this chapter.

Amanda Webber has an MSc in primate conservation and a PhD in anthropology from Oxford Brookes University. She is a Lecturer in the Conservation Science Department of Bristol Zoological Society, working with UWE Bristol and University of Bristol students. Her research interests are human-wildlife interactions, perceptions of animal species (particularly 'problem' animals) and wildlife conservation. She is currently developing several research projects in Madagascar examining people-wildlife interactions.

Stewart Thompson is Professor of Biodiversity Conservation in the Department of Biological and Medical Sciences, Oxford Brookes University. He is the director of a research cluster whose work explores the linkages between wildlife protection mechanisms/policies and landscape scale ecology; the effects of land-use change on wildlife; threatened species conservation in developing countries and wildlife tourism and human-wildlife conflict resolution. Many of his research interests are underpinned by the use of Geographic Information Systems as data repositories and predictive management tools. He has a particular interest in threatened species conservation, with several ongoing projects in developing countries that seek to enhance endangered species population numbers and distribution patterns.

Neil Bailey has an academic background in the use of GIS and landscape ecology for conservation benefit. Twelve years ago, Neil started investigating how GPS-enabled devices could be used by non-experts to gather environmental data to feed into GIS and better inform environmental management. This led to the formation of an Oxford Brookes University spin-out (WildKnowledge), with his colleague Stewart Thompson. WildKnowledge engages the public in exploring and recording the environment through a suite of mobile apps. As a result of this experience, Neil developed an interest in the 'science behind citizen science' and has accumulated expertise in how to recruit for, and maintain interest in, citizen science projects. Neil is based at Earthwatch Europe, an international NGO that specializes in citizen science as a vehicle for engaging people in environmental issues.

Nancy Priston is an Honorary Research Associate at Oxford Brookes University. Her research examines human-wildlife conflict with a predominantly interdisciplinary approach, incorporating both the perspectives of wildlife and local people.

References

African Elephant Specialist Group. 2016. 'African Elephant Database.' Retrieved 25 May 2016 from www.elephantdatabase.org.

Ahmadi, M., et al. 2013. 'A Predictive Spatial Model for Gray Wolf (*Canis lupus*) Denning Sites in a Human-Dominated Landscape in Western Iran', *Ecological Research* 28: 513–521.

Anderson, K., and K.J. Gaston. 2013. 'Lightweight Unmanned Aerial Vehicles will Revolutionise Spatial Ecology', *Frontiers in Ecology and the Environment* 11(3): 138–146.

Behdarvand, N., et al. 2014. 'Spatial Risk Model and Mitigation Implications for Wolf-Human Conflict in a Highly Modified Agroecosystem in Western Iran', *Biological Conservation* 177: 156–164.

Boyce, M.S., et al. 2010. 'Temporal Autocorrelation Functions for Movement Rates from Global Positioning System Radio Telemetry Data', *Philosophical Transactions of the Royal Society B* 365: 2213–2219.

Brown, G.G., and P. Reed. 2009. 'Public Participation GIS: A New Method for Use in National Forest Planning', *Forest Science* 55(2): 166–182.

Brown, G., and D. Weber. 2011. 'Public Participation GIS: A New Method for National Park Planning', *Landscape and Urban Planning* 102: 1–15.

Bryan, B.A., et al. 2010. 'Comparing Spatially Explicit Ecological and Social Values for Natural Areas to Identify Effective Conservation Strategies', *Conservation Biology* 25(1): 172–178.

Carver, S. 2003. 'The Future of Participatory Approaches Using Geographic Information: Developing a Research Agenda for the 21st Century', *URISA* 15(1): 61–71.

Chambers, R. 2006. 'Participatory Mapping and Geographic Information Systems: Whose Map? Who is Empowered and Who Disempowered? Who Gains and Who Loses?' *EJISDC* 25(2): 1–11.

Cinderby, S. 1999. 'Geographic Information Systems for Participation: the Future of Environmental GIS?' *International Journal of Environment and Pollution* 11(3): 304–315.

Cox, C., et al. 2014. 'Applying Public Participation Geographic Information Systems to Wildlife Management', *Human Dimensions of Wildlife* 19(2): 200–214.

Cress, D., and Z. Zommers. 2014. 'Emerging Technologies: Smarter Ways to Fight Wildlife Crime', UNEP *Global Environment Alert Service Bulletin*. Retrieved 27 June 2016 from http://na.unep.net/geas/getUNEPPageWithArticleIDScript.php?article_id=113.

Douglas-Hamilton, I., T. Krink and F. Vollrath. 2005. 'Movements and Corridors of African Elephants in Relation to Protected Areas', *Naturwissenschaften* 92: 158–163.

Elwood, S. 2006. 'Critical Issues in Participatory GIS: Deconstructions, Reconstructions and New Research Directions', *Transactions in GIS* 10(5): 693–708.

Enari, H., and T. Suzuki. 2010. 'Risk of Agricultural and Property Damage Associated with the Recovery of Japanese Monkey Populations', *Landscape and Urban Planning* 97: 83–91.

Fox, J., P. Yonzon and N. Podger. 1996. 'Mapping Conflicts between Biodiversity and Human Needs in Langtang National Park, Nepal', *Conservation Biology* 10(2): 562–569.

Frair, J.L., et al. 2010. 'Resolving Issues of Imprecise and Habitat-Biased Locations in Ecological Analyses Using GPS Telemetry Data', *Philosophical Transactions of the Royal Society B* 365: 2187–2200.

Gautier, J.-P., and S. Biquand. 1994. 'Primate Commensalism', *Revue d'Ecologie (La Terre et La Vie)* 49: 210–212.

Graham, M.D., et al. 2009. 'The Movement of African Elephants in a Human-Dominated Land Use Mosaic', *Animal Conservation* 12: 445–455.

GRASS GIS. 2016. Geographic Resources Analysis Support System. Retrieved 1 June 2016 from https://grass.osgeo.org.

Haslett, J.R. 1990. 'Geographic Information Systems: A New Approach to Habitat Definition and the Study of Distributions', *Trends in Ecology and Evolution* 5(7): 214–218.

Hebblewhite, M., and D.T. Haydon. 2010. 'Distinguishing Technology from Biology: A Critical Review of the Use of GPS Telemetry Data in Ecology', *Philosophical Transactions of the Royal Society B* 365: 2303–2312.

Hill, C.M. 1997. 'Crop Raiding by Wild Vertebrates: The Farmer's Perspective in an Agricultural Community in Western Uganda', *International Journal of Pest Management* 43(1): 77–84.

Hill, C.M., F.V. Osborn and A.J. Plumptre. 2002. *Human–Wildlife Conflict: Identifying the Problem and Possible Solutions*. Albertine Rift Technical Report Series No. 1: Wildlife Conservation Society.

Hodgson, D.L., and R.A. Schroeder. 2002. 'Dilemmas of Counter-Mapping Community Resources in Tanzania', *Development & Change* 33: 79–100.

Hoffman, T.S., and M.J. O'Riain. 2012. 'Landscape Requirements of a Primate Population in a Human-Dominated Environment', *Frontiers in Zoology* 9(1). doi:10.1186/1742-9994-9-1.

Hughes, K. 2003. 'The Global Positioning System, Geographical Information Systems and Remote Sensing', in J.M. Setchell and D.J. Curtis (eds), *Field and Laboratory Methods in Primatology: a Practical Guide*. Cambridge: Cambridge University Press, pp. 57–73.

Humle, T., et al. 2014. 'Biology's Drones: Undermined by Fear', *Science* 344(6190): 1351.

Jachowski, D.S., R. Slotow and J.J. Millspaugh. 2014. 'Good Virtual Fences Make Good Neighbours: Opportunities for Conservation', *Animal Conservation* 17: 187–196.

Kahler, J.S., G.J. Roloff and M.L. Gore. 2012. 'Poaching Risks in Community-Based Natural Resource Management', *Conservation Biology* 27(1): 177–186.

Karanth, K.K., et al. 2012. 'Assessing Patterns of Human-Wildlife Conflicts and Compensation Around a Central Indian Protected Area', *PLoS ONE* 7(12): e50433. doi:10.1371/journal.pone.0050433.

King, L.E., I. Douglas-Hamilton and F. Vollrath. 2011. 'Beehive Fences as Effective Deterrents for Crop-Raiding Elephants: Field Trials in Northern Kenya', *African Journal of Ecology* 49(4): 431–439.

Koenig, W.D. 1999. 'Spatial Autocorrelation of Ecological Phenomena', *Trends in Ecology and Evolution* 14(1): 22–26.

Kretser, H.E., P.D. Curtis and B.A. Knuth. 2009. 'Landscape, Social and Spatial Influences on Perceptions of Human-Black Bear Interactions in the Adirondack Park, NY', *Human Dimensions of Wildlife* 14(6): 393–406.

Le Lay, G., P. Clergeau and L. Hubert-Moy. 2001. 'Computerized Map of Risk to Manage Wildlife Species in Urban Areas', *Environmental Management* 27(3): 451–461.

Lewis, D.M. 1995. 'Importance of GIS to Community-Based Management of Wildlife: Lessons from Zambia', *Ecological Applications* 5(4): 861–871.

Malo, J.E., F. Suárez and A. Diez. 2004. 'Can We Mitigate Animal-Vehicle Accidents Using Predictive Models?' *Journal of Applied Ecology* 41: 701–710.

Marshall, J. 2007. 'Smartphones are the PCs of Developing World', *New Scientist* 195(2615): 24–25.

McLennan, M.R. and C.M. Hill. 2012. 'Troublesome Neighbours: Changing attitudes towards Chimpanzees (*Pan troglodytes*) in a Human-Dominated Landscape in Uganda', *Journal for Nature Conservation* 20(4): 219–227.

Michelmore, F. 1994. 'Keeping Elephants on the Map: Case Studies of the Application of GIS for Conservation', in R.I. Miller (ed.), *Mapping the Diversity of Nature*. London: Chapman & Hall, pp. 107–125.

Mulero-Pázmány, M., et el. 2014. 'Remotely Piloted Aircraft Systems as a Rhinoceros Anti-Poaching Tool in Africa', *PLoS ONE* 9(1): e83873. doi:10.1371/journal.pone.0083873.

Mwavu, E.N., and E.T.F. Witkowski. 2008. 'Land-Use and Cover Changes (1988–2002) Around Budungo Forest Reserve, NW Uganda: Implications for Forest and Woodland Sustainability', *Land Degradation and Development* 19: 606–622.

Pigott, R. 2014. 'Heathrow Plane in Near Miss with Drone', BBC News. Retrieved 2 June 2016 from http://www.bbc.co.uk/news/uk-30369701.

Rabinowitz, A., and K.A. Zeller. 2010. 'A Range-Wide Model of Landscape Connectivity and Conservation for the Jaguar, *Panthera onca*', *Biological Conservation* 143: 939–945.

Rambaldi. G. 2010. *Participatory Three-Dimensional Modelling: Guiding Principles and Applications*, 2010 edition. Wageningen, The Netherlands: The Technical Centre for Agricultural and Rural Cooperation.

Randy-Gimblett, H. 2002. 'Integrating Geographic Information Systems and Agent-Based Technologies for Modeling and Simulating Social and Ecological Phenomena', in H. Randy-Gimblett (ed.), *Integrating Geographic Information Systems and Agent-Based Modeling Techniques for Simulating Social and Ecological Processes*. Oxford: Oxford University Press, pp. 1–20.

Save the Elephants. 2016. 'Geo Fencing.' Retrieved 2 June 2016 from www.savetheelephants.org/geo-fencing.

Schiffman, R. 2014. 'Drones Flying High as New Tool for Field Biologists', *Science* 344: 459.

Sieber, R. 2006. 'Public Participation Geographic Information Systems: a Literature Review and Framework', *Annals of the Association of American Geographers* 96(3): 491–507.

Sitati, N.W., et al. 2003. 'Predicting Spatial Aspects of Human-Elephant Conflict', *Journal of Applied Ecology* 40: 667–677.

Smith, R.J., and S.M. Kasiki. 2000. *A Spatial Analysis of Human-Elephant Conflict in the Tsavo Ecosystem, Kenya*. A report to the African Elephant Specialist Group, Human-Elephant Conflict Task Force, IUCN, Gland, Switzerland.

Space for Giants. 2014. 'Tweeting Elephants – Behind the Scenes', *The Independent*. Retrieved 2 June 2016 from www.independent.co.uk/voices/comment/tweeting-elephants-behind-the-scenes-9074896.

Swetnam, R.D., and B. Reyers. 2011. 'Meeting the Challenge of Conserving Africa's Biodiversity: the Rof GIS, Now and in the Future', *Landscape and Urban Planning* 100: 411–414.

Tenenbaum, N. 2012/13. 'Into the Wild', *Unmanned Vehicles* 17(6): 27–29.

Thirgood, S., R. Woodroffe and A. Rabinowitz. 2005. 'The Impact of Human-Wildlife Conflict on Human Lives and Livelihoods', in R. Woodroffe, S. Thirgood and A. Rabinowitz (eds), *People and Wildlife: Conflict or Coexistence?* Cambridge: Cambridge University Press, pp. 13–26.

Tomkiewicz, S.M., et al. 2010. 'Global Positioning System and Associated Technologies in Animal Behaviour and Ecological Research', *Philosophical Transactions of the Royal Society B* 365: 2163–2176.
Treves, A., et al. 2004. 'Predicting Human-Carnivore Conflict: a Spatial Model Derived from 25 Years of Data on Wolf Predation on Livestock', *Conservation Biology* 18(1): 114–125.
Treves, A., et al. 2011. 'Forecasting Environmental Hazards and the Application of Risk Maps to Predator Attacks on Livestock', *BioScience* 61(6): 451–458.
Treves, A., and K.U. Karanth. 2003. 'Human-Carnivore Conflict and Perspectives on Carnivore Management Worldwide', *Conservation Biology* 17(6): 1491–1499.
Tumusiime, D.M., and H. Svarstad. 2011. 'A Local Counter-Narrative on the Conservation of Mountain Gorillas', *Forum for Development Studies* 38(3): 239–265.
Tweheyo, M., C.M. Hill and J. Obua. 2005. 'Patterns of Crop Raiding by Primates around the Budongo Forest Reserve, Uganda', *Wildlife Biology* 11(3): 237–247.
Waldron, J.L., et al. 2013. 'Using Occupancy Models to Examine Human-Wildlife Interactions', *Human Dimensions of Wildlife* 18(2): 138–151.
Wall, J., et al. 2014. 'Novel Opportunities for Wildlife Conservation and Research with Real-Time Monitoring', *Ecological Applications* 24(4): 593–601.
Webber, A.D. 2006. 'Primate Crop Raiding in Uganda: Actual and Perceived Risks around Budongo Forest Reserve', PhD thesis, Oxford Brookes University, Oxford.
Webber, A.D., and C. M. Hill. 2014. 'Using Participatory Risk Mapping (PRM) to Identify and Understand People's Perceptions of Crop Loss to Animals in Uganda', *PLoS ONE* 9(7): e102912. doi:10.1371/journal.pone.0102912.
WildKnowledge. 2011. 'Lions and Warriors.' Retrieved 3 January 2015 from www.wildknowledge.co.uk/news/2011/11/lions-and-warriors/.
Wilson, S.M., et al. 2005. 'Natural Landscape Features, Human-Related Attractants, and Conflict Hotspots: a Spatial Analysis of Human–Grizzly Bear Conflicts', *Ursus* 16(1): 117–129.
———. 2006. 'Landscape Conditions Predisposing Grizzly Bears to Conflicts on Private Agricultural Lands in the Western USA', *Biological Conservation* 130: 47–59.
Wilson, S.M., G.A. Neudecker and J. J. Jonkel. 2014. 'Human–Grizzly Bear Coexistence in the Blackfoot River Watershed, Montana: Getting Ahead of the Conflict Curve', in S.G. Clark and M.B. Rutherford (eds), *Large Carnivore Conservation: Integrating Science and Policy in the North American West*. Chicago, IL, and London: University of Chicago Press, pp. 177–214.
Woodroffe, R., S. Thirgood and A. Rabinowitz. 2005. *People and Wildlife: Conflict or Coexistence?* Cambridge: Cambridge University Press.
Wright, H.L., I.R. Lake and P.M. Dolman. 2012. 'Agriculture – a Key Element for Conservation in the Developing World', *Conservation Letters* 5(1): 11–19.

Index

Africa, 24, 65, 129, 141, 170, 201
  East Africa, 19, 20t1.1, 24, 25t1.3, 141
  West Africa, 19, 21
agriculture, 5, 21, 26, 37, 49, 51, 52, 54, 61, 117, 128, 129, 170, 174, 198
  agricultural expansion, 24
Amazonian societies, 95
animal, 2, 4, 7, 15, 57, 68, 72, 96, 97, 98
  'bad', 6, 75, 79–82, 85
  'charismatic', 83
  fear of humans, 38, 40–42
  'good', 6, 75, 76–79, 82
  habituation to crop protection, 184
  habituation to people, 38, 41, 42, 44–45, 46
  personhood, 6–7, 97–101
  social constructions of, 6, 7, 8, 21f1.2, 22, 69, 85, 136, 137, 144n2
  symbols, 2, 5–6, 7, 24, 50, 54, 77, 84, 85f4.5, 148, 155, 200
  transforming animals from wild to non-wild, 60–61
  vermin, 24, 26, 69, 71, 76, 80, 81, 83, 84, 131, 142, 143 (*See also* 'pest')
  welfare, 60, 66, 69, 84, 110
Animals of Farthing Wood (Dann), 72
animism, 99
attitudes, 3, 8, 15, 16, 18, 28, 108–110, 121, 128, 151
  behaviour, and, 108–110
  biodiversity conservation, and, 15, 18–19, 19–20, 21–23, 112
  changes in, 8–9, 100, 117, 121, 128, 129, 165
  protected areas, 15, 18–23
  stability of, 128, 129, 131, 132, 142–43
  underlying factors, 15, 20, 20t1.1, 108–109
  urban versus rural views, 5, 8
  wildlife, 111, 112, 114, 115, 122

baboon (*Papio anubis*), 129, 131, 136
  attitudes to, 9, 10
  crop pest, as, 129, 131, 136, 136t7.1, 138, 140, 142, 178, 184, 198, 204
  hunting of, 140, 142
  'worst' animal, 137, 138, 139, 144
badger (*Meles meles*), 66, 67, 76, 80, 82
  anti-cull, 67f4.1
  'bad', 6, 75, 79–82, 85
  Badgers Act, 1973, the, 71
  BadgerBadgerBadger (internet craze), 73, 87n9
  badger watching, 76, 77, 87n6
  baiting and hunting, 66, 68, 69, 75, 76
  contemporary bovine TB debates, 75–82, 83
  corporate logos, 72, 73, 87n5
  crop damage, 80, 83
  culling, 6, 15, 66–67, 68, 75. 78, 79, 81, 82, 83, 85, 87n10
  cultural significance in the UK, 68–75, 83, 87n3
  disease transmission, 66, 68, 79, 80, 83
  farming community's attitudes to, 77
  'good', 6, 75, 76–79, 82
  heraldry, 72, 72f4.2, 73, 77
  historical framing of, 68–75
  impact on farmers, 65, 70
  in literature, 71–75, 80
  mass media, 6, 69–71, 75–82

Index

negative representation of, 73–74
positive representations of, 69–70
predator, 70, 74, 75, 80, 81
protected pest, 6
protected status, 12n2, 67, 81, 86
Protection of Badgers Act, 1992, 12n2
social constructions of, 6, 68–82
symbolic nature of, 77, 84, 85f4.5
'vermin', as, 69, 70, 71, 81, 83, 84
victim, as, 78, 79
Badger Consultative Panel, 73
BBC Natural History Unit, 73, 75, 84
bear
   black (*Ursus americanus*), 112, 113t6.2, 121, 196
   brown (*Ursus arctos*), 55, 58, 129
   grizzly (*Ursus arctos horribilis*), 69, 196, 197, 199–200
beliefs, 28, 109, 110, 112, 155
   environment, 18
   spiritual, 28, 155
   wildlife, 21, 21f1.2, 27, 114, 119
benefits of conservation, 17, 18, 19, 22
   costs and benefits analysis, 16, 17f1.1, 18, 20, 183
   from wildlife, 26
biodiversity conservation, 15, 16, 20, 62, 97, 103, 194
   planning, 196, 197
biosocial approach, 2–3, 4–9
   value of, 3, 4, 11
bovine TB, 6, 66–67
   and badgers, 6, 66–67, 68, 75–82
   and domestic cattle, 65, 66, 68, 75, 76, 79, 80, 86
   debates in the UK, 65–86
   economic impacts, 66
   health risks, 66, 83, 85
   history of in the UK, 66
   public scientific controversy, 66
Britain, rural life, 75, 81, 82, 83
Budongo Forest Reserve, Uganda, 174, 175f9.1, 198–99
bushmeat, 19, 21, 22, 101, 142

capacity building, 162, 163
carnivores, 5, 50, 51, 55, 57, 58, 59, 61, 96, 128, 195
   African wild dog (*Lycaon pictus*), 163
   attitudes to, 55, 60, 128

black bear (*See under* bear)
brown bear (*See under* bear)
conflicts, 51, 62, 199–200
hunting quotas, 55
lynx (*Lynx lynx*), 51, 55, 63n2
management of, 55–61, 56f3.2, 62, 152
mountain lion (*Puma concolor*), 120
wolf (*See* wolves)
wolverine (*Gulo gulo*), 51, 55
chimpanzee (*Pan troglodytes*), 6, 19, 138
   crop damage, 131, 136, 136t7.1, 137, 140, 178
   flagship species, 21
   perceptions of, 21–22
   pests, 25t1.3, 138, 139
Christianity among the Trio, 95, 101–2, 103
CITES (Convention on International Trade in Endangered Species of Wild Fauna and Flora), 16, 25
Clare, John, 71, 87n2
   The Badger, 71
class politics, 75
coexistence with wildlife. *See* human-wildlife coexistence
Cognitive Hierarchy Framework, 8, 108–9, 121
community-based biodiversity conservation, 27–28, 29
compensation, 19, 50, 55, 61, 96, 128, 150, 154, 156, 158, 163, 173
   and GIS, 196
complex problems, 1, 9
conflicts about wildlife, 103–4, 149–150
   automobile collisions, 117
   badger *(See* badger)
   baboon (*See* baboon)
   carnivores, 51, 62, 199–200
   causes of, 15–18, 37–38, 49
   crop damage (*See* crop damage)
   elephant (*See* elephant)
   'hotspots', 113, 127, 129, 141, 142, 143, 195–96, 199
   Japanese macaque (*Macaca fuscata*), 36–46
   livestock losses, 51–56, 58, 119, 129, 153, 156
   perceptions of, 16, 17, 21–22, 28, 196
   predation, 74, 83, 129, 162, 197

retaliatory killing, 3, 24
rural depopulation, 5, 37–38, 41, 52
terminology, 1, 2, 3–4, 6, 15–16, 148, 171, 187n3
Conflict Intervention Triangle, 153, 157–60, 158f8.2
conflict mitigation, 1, 2, 3, 9–11, 29, 50, 55–60, 104n5, 171–72, 199–201
    culling 'problem' animals, 45, 47n6, 51, 58, 173, 201 (*See also under* badger)
    decision-making process, 156, 158–60, 164–65, 184–85, 194–95
    lethal control, 110, 112, 172
    public acceptance, 8, 108, 112–14, 119–20, 120f6.5, 121
    reconciliation, 156, 157, 162–63
    resolution, 19, 86, 117, 120–21, 148, 149, 154f8.1, 157, 159
    stakeholder engagement, 172, 174–77, 184–86, 195–96, 197–98
    unintended consequences of, 61–63
    See also crop protection; livestock protection
conflict models, 148, 153–160, 154f8.1
    dispute level, 153, 154, 155, 156, 158, 164
    identity conflict level, 154–55, 156–57, 164
    underlying conflict level, 150, 153–154, 156–157, 159
    See also Conflict Intervention Triangle
conflict narratives, 7–8, 9
Conservation Conflict Transformation (CCT), 9–10, 148–66
    participatory process, 9, 157, 170–86
Conservation International (CI), 103
Cosslett, Tess, 73
counterhabituation, 40–42
crop damage, 3, 4, 9, 22, 104n1, 127, 130, 131, 135, 136t7.1, 137, 140, 142, 143, 171, 196, 202
    badger, 80, 83
    complaints about, 9, 11, 26, 131, 132, 140, 142, 143, 144n2, 156, 173, 177, 178
    domestic animals, 24, 104, 131, 176, 177, 185
    elephants, 19, 25–27, 129, 131, 137, 141, 143
    impact on livelihoods, 171, 173, 174, 183
    measurement of, 127, 131, 171
    patterns of, 131, 133, 143, 180, 198–99
    perceptions of risk, 9, 24, 127, 128, 129, 133, 141, 143, 174, 196
    perceptions of wildlife, 128, 177
    primates, 4–5, 19, 21–22, 36, 37–38, 43, 44, 45, 46, 131, 136t7.1, 137, 178, 198–199
    risk factors, 129, 133, 141, 172, 186
crop loss. *See* crop damage
crop protection, 9–11, 38–45, 140–41, 170, 174
    beehives, 27
    chemical repellents, 10, 172, 184, 187n2
    chilli pepper, 27, 152, 181, 182t9.1, 183–84
    dogs, 40, 47n3, 181, 182t9.1
    drones, 203
    early warning systems, 10, 27, 180, 199, 200
    electric fencing, 39, 195
    fencing, 10, 27, 38–39, 40, 42, 45, 140, 152, 156, 165, 172, 173, 178, 179, 181, 182t9.1, 183–85, 195–96, 200–201, 203
    GIS, 196–97, 198–99
    guarding fields, 39–40, 46, 131, 140, 172, 173, 178–79, 181–85, 182t9.1, 199
    lethal control, 9, 45, 47n6, 131, 140, 142, 172, 173, 198, 200
    provisioning, 43–45
    psychological barriers, 42
    scarecrow, 40–41, 178, 179 (*See also kakashi*)
    solar lights, 181, 182t9.1, 183, 185
    traditional methods, 128, 140–41, 173–74, 177–78
    trenches, 140, 181, 182t9.1
crop protection tools, development of, 170–74, 179–84
crop protection tools, evaluation of, 179–84, 182t9.1
crop-raiding. *See* crop damage
crop-raiding event (CRE), 176–77, 178, 187n3

cultural identity, 148
cultural landscape, 54, 96, 63n3
cultural value of nature, 54

Daily Mail, 79, 81
Daily Telegraph, 78, 81
damage to property, 1, 2, 6, 11, 38, 150, 196, 200
    primates entering houses, 38
Dann, Colin, 72
    Animals of Farthing Wood, 72
decision-making process, 156–60
    exclusion from, 16, 18, 154, 159
    inclusion in, 158, 160, 164
deforestation and forest degradation, impacts of, 21, 22, 45, 102, 104n5
deterrents. *See* crop protection
domestic animals, 10, 21, 37, 68, 96, 102, 135, 185
    cattle, 66, 68, 77, 79, 102, 149, 157, 135
    crop damage, 24, 104, 131, 176, 177, 185
    livestock losses, 3, 10, 55, 96, 149, 157, 162, 163, 197
'*dorobō*' (monkey thieves), 37
drivers of conflict, 28, 37–38, 128, 129, 135, 148–49, 164–65
    social factors, 22–23, 27–28, 49–50, 51–5, 59, 28, 121, 151, 153–60, 154f8.1
drones, 203

economic impact, 5, 18, 20, 22, 49, s50–51, 52, 55, 57, 61, 66, 96, 134, 150, 151, 180
    Wildlife Value Orientations, 144, 122, 124n3
education, 15, 22, 29, 156, 163, 164, 202
    attitudes to, 5, 20t1.1, 114, 156
    attitudes to environment, 54
    Wildlife Value Orientations, 110, 11t6.1, 113, 114, 115, 116f6.2
elephant (*Loxodonta africana and Elephas maximus spp.*), 19, 29, 200–201
    crop damage by, 19, 25–27, 129, 131, 137, 141, 143
    electric fencing, 152, 195–96, 200
    flagship species, 24

human-elephant conflict (HEC), 20, 24, 25–26, 27, 204
    lethal control, 200
    perceptions of, 24–27
    symbolism, 24, 200
elephant-proof fencing, 27
Engai or 'monkey damage', 36–38, 45–46

farmer
    engagement, 10–11, 170–87
    livestock, 50, 57, 62 (*See also* livestock)
    responses to 'conflict', 7–8, 26, 38–39, 45, 57–58, 59
    views on wildlife, 3, 9, 22, 41, 43, 173, 174, 178
fishing, 22, 112
flagship species, 21, 24
forest industries, 5, 37, 51, 22, 54
foxhunting, 6, 70, 75, 81

gender, 15, 16, 17, 19, 20t1.1, 28, 77, 155, 173
Geographic Information Systems (GIS), 11, 195
    animal behaviour, 200
    encouraging stakeholder engagement, 11, 197–98, 199–200, 201
    future developments, 202–3, 204
    ground truthing, 197, 198, 201
    identifying 'conflict hotspots', 11, 199
    limitations of, 201–2
    participatory GIS (PGIS), 197–98, 199–200
    participatory, 3D Modelling (P3DM), 198
    planning tool, 196, 197, 199
    predicting conflict hotspots, 196–97, 197, 198–99
    understanding patterns of primate crop damage, 198–99
    visualising human-wildlife interactions, 11, 195–201
Global Positioning System (GPS), 143, 195, 201
governance, 17, 18, 19, 68
Grahame, Kenneth, 71, 73, 84
    The Wind in the Willows, 71, 76, 80, 82, 84
Guardian, The, 77, 80

Guinea Bissau, 19, 21–23

Harry Potter (Rowling), 72
herding, 51, 58, 59. *See also under* livestock
herding instinct, 51
human-animal relationship, 6, 7, 17, 69, 97, 98–101
human-elephant conflict (HEC), 20, 24, 25–26, 27, 204
human-human conflict, 3, 4, 10, 45, 46, 59, 61–62
human injury or death, 20, 20t1.1, 99, 100
human-wildlife coexistence, 2, 3, 9, 16, 29, 77, 86, 149, 150, 151, 152, 153, 159, 160, 163, 164, 165, 194, 200
human-wildlife conflict. *See* conflicts about wildlife
Human-Wildlife Conflict Collaboration (HWCC), 97, 150, 153, 156, 157, 165, 166n1
human-wildlife conflict mitigation. *See* conflict mitigation
hunting and trapping, 21, 45, 70, 75, 112, 117, 140, 161, 172
  hunters, 38, 40, 50, 57, 59–60, 62, 102, 111t6.1, 140, 142
  illegal, 16, 50, 62, 78, 82, 131, 152
  recreational, 20t1.1, 112, 113t6.2, 117, 124n3

identity, 18, 20, 28, 72, 77, 98, 100, 150, 151, 152
  cultural identity, 148, 150
  identity conflict, 154–57, 158, 163, 200
incentives, 49, 150, 173
  economic, 17, 17f1.1, 19, 49
Independent, The, 76, 80
India, 21, 26, 63n3, 197
International Union for Conservation of Nature (IUCN), 4

Jacques, Brian, 72
  Redwall series, 72
Japan, 4, 5, 36–46, 47n5, 51
Japanese macaque *(Macaca fuscata)*, 36–46, 47n5, 196

kakashi, 40–41
Kenya, 23t1.2, 24, 25t1.3, 141, 163, 195, 198, 200, 202
Kenya Wildlife Service, 24
Kibale National Park, Uganda, 25t1.3, 127–43, 144n1

Lewis, C.S., 72
  Prince Caspian, 72
limitations of current conservation practice, 150–53
livelihoods, 27, 28, 29, 54, 102, 170, 179
  negative impact on, 15, 18, 22, 27, 43, 58, 65, 121, 170–72, 174, 181
  positive contribution to, 44, 77, 176, 200
livestock
  herder, 7, 50, 59 (*See also* herding)
  losses, 51–56, 58, 119, 129, 153, 156, 162
livestock protection
  conflict management, 163, 199–200
  culling carnivores, 50, 55, 56, 58, 65, 66–67, 68, 162
  electric fences, 51, 57–60, 63n2
  guard dogs, 51
  impact on wildlife, 58–59

Maslow's 'hierarchy of needs', 114, 150
media influence, 6, 17f1.1, 29, 58, 59
middle-class, 4, 54, 62, 75
modernization, 52, 108, 114–17, 122
monkey chasing. *See* Oiharai
monkey parks, 43–45, 45–46, 47n5
monkey thieves. *See* '*dorobō*'
mountain lion *(Puma concolor)*, 120

natural resource use, 5, 22, 49, 50, 51, 96, 103, 104n3
  economic decline, 97, 103
  human-animal resource competition, 83, 96, 104, 117, 170, 194, 196, 201
  restrictions on, 15, 19, 54, 57
nature and culture boundaries, 95, 97, 103
Neal, Ernest, 72
  The Badger, 72, 73
Norway, 4, 5, 6, 49–62, 63n3

Observer magazine, 76
Oiharai (monkey chasing), 41, 45

pastoralists, 20, 25
    Maasai, 24
    See also under livestock
peace-building, 149, 150, 153
perceptions, 8–9, 15–29, 128
    change over time, 127–43
    of conservation initiatives, 18–19, 21–27
    of protected areas, 19–20, 20t1.1A, 21
personhood, 6–7, 150
    among the Trio, 7, 97–101
'pest', 3, 9, 16, 20t1.1B, 26, 28, 29, 40, 41, 42, 56, 83, 85, 128
    label, 3, 6 (*See also* terminology)
    'protected' pest, 6
    ranking of, 24, 25t1.3, 138–39, 142, 144n2, 177
    social construction of, 5, 6, 84
pestilence discourse, 83, 84
poaching, 7, 21, 131, 141, 153, 203
    anti-poaching, 160–61, 201
political action, 7–8, 117, 160–162
politics, 44, 51, 81, 86, 148, 150
    conflict, of, 4, 15–29, 55, 61, 103
    environmental politics, 72
    social class, 5, 50, 52, 54, 59, 62, 75, 81 (*See also* middle class; working class)
    wildlife management, 55–57, 59–60
porcupine (*Hystrix cristata*), 25, 176
Potter, Beatrix, 73, 74f4.4, 82
    The Tale of Mr Tod, 73, 74, 74f4.4, 79
power imbalance, 2, 7, 104, 148, 155, 159
power relationships, 3, 6, 16, 17f1.1, 50, 61, 81
predator, 4, 6, 8, 24, 36, 50, 55, 58, 99, 100, 121
    conservation project, 163
    history of predator removal, 117 (*See also* badger; wolves)
    threat to human safety, 110
predators
    African wild dog (*Lycaon pictus*), 163
    lynx (*Lynx lynx*), 51, 55, 63n2
    mountain lion (*Puma concolor*), 120
    wolverine (*Gulo gulo*), 51, 55
    See also badger; bear; wolves

primates, 21, 22, 24, 25t1.3, 36, 129, 142, 177, 178, 181, 182t9.1, 184, 186, 198, 199, 202
    baboon (*See* baboon)
    black and white colobus monkey (*Colobus guereza occidentalis*), 178
    blue monkey (*Cercopithecus mitis stuhlmanni*), 178
    chimpanzee (*See* chimpanzee)
    Japanese macaque (*See* Japanese macaque)
    red-tailed monkey (*Cercopithecus ascanius schmidti*), 129, 131, 136–40, 142, 178, 198
    spider monkey (*Ateles* sp.), 100, 102
    vervet monkey (*Chlorocebus aethiops*), 25t1.3, 131, 142, 178
Prince Caspian (Lewis), 72
Private Eye, 76
problem animals, 6, 16, 24, 56, 127. *See also* 'pest'
protected areas, 15, 16, 19–20, 20t1.1, 21, 24, 26, 28, 54, 101, 152, 200, 203
    Amboseli National Park, Kenya, 24, 29
    Budongo Forest Reserve, Uganda, 174, 175f9.1, 198–99
    Kibale National Park, Uganda, 25t1.3, 127–43, 144n1
    nature reserves, 97
    Selous Game Reserve, Tanzania, 19–20, 29
    Tombali and Lagoas de Cufada Natural Park, Guinea Bissau, 21, 23t1.2A
public opinion, 10, 60, 61, 110, 114, 117, 197
public protest, 26, 117
public support, 8, 19, 44, 45, 59, 107, 108, 110, 112, 113–14, 119, 151, 161
    lack of, 19, 50, 109, 121, 127
    supporters of wolves, 60–61, 119

rancher, 25, 153, 161–62, 199–200
Randomised Badger Culling Trial (RBCT), 66
recreation, 45, 54, 111, 112, 117
    birdwatching, 87n7, 112, 117
    recreational hunting, 20t1.1, 112, 113t6.2, 117, 119, 124n3

Index 217

wildlife tourism, 22, 17
wildlife viewing, 76, 87n6, 110, 111, 112, 117
REDD+, 102, 103, 104n5
Redwall series (Jacques), 72
research methods
   attitudinal measures, 110–11
   cross-sectional data, 129
   developing crop protection tools, 175–77
   framing analysis, 75
   grounded theory, 75
   longitudinal data, 129
   methodological challenges, 142–43
retaliatory killing, 3, 24
revenue-sharing, 19, 24
re-wilding, 54
risk
   animal body size, 24
   health risks, 23
   livelihood, 23t1.2, 26, 28, 43, 58, 65, 121, 171, 174, 176
   narratives of, 53, 84
   risk mapping, 22, 23t1.2, 24
   risk maps, 11, 26, 195–96
   See also badger; crop damage
Rowling, J.K., 72
   Harry Potter, 72
rural depopulation, 5, 37–38, 41, 52

satoyama (village forest), 37, 44
Scott, James, 7–8
   'Weapons of the Weak', 7–8, 161–62
social carrying capacity for wildlife, 11, 149, 150, 152, 165
social change, 5, 16, 49–50, 52, 134–35, 165
social class, 59. *See also* middle-class; working class
social conflicts, 2, 9, 10, 52, 108, 110, 111–14, 118, 119, 148, 150, 151, 155–56, 197, 198
   transforming social conflicts, 149, 152, 154, 156, 163–64
social psychology, 8, 121, 150
socioeconomic status, 8, 114, 122, 128
socio-zoological scale, 21, 21f1.2, 27
South America, 21, 104
spirit attack, 99, 100
spirit-master, 98–99, 100, 101

Sprague, David, 37
Springwatch (TV series), 75
stakeholder engagement, 164, 172, 173, 174–75, 181, 184–86, 197–98
Suriname, 7, 95–104
symbolism, 2, 5–6, 7, 24, 50, 54, 77, 84, 85f4.5, 148, 155, 200

Tanzania, 19–20, 198
terminology, 2, 3–4, 6, 15, 29, 156, 171
   'why labels are important', 3–4, 15–16, 148
The Badger (Neal), 72, 73
The Badger (Clare), 71
The Combe (Thomas), 71, 84
The Sword in the Stone (White), 72
The Tale of Mr Tod (Potter), 73, 74, 74f4.4, 79
The Wind in the Willows (Grahame), 71, 76, 80, 82, 84
Thomas, Edward, 71, 77
   The Combe, 71, 84
threat to human safety, 2, 6, 110, 111, 112, 119, 200
Times, The, 69, 70, 71, 73, 77, 78, 79, 81
tolerance for wildlife, 9, 16, 142, 171
   and attitudes, 8, 120
   policy of zero tolerance, 41, 42
   promoting tolerance, 9–10, 28, 29, 165
   social carrying capacity, 11, 149, 150, 152, 165
tourism, 17, 20, 22, 43–44, 46, 52, 54, 185
Trio of Suriname, 95–104
trust building, 163, 185, 197
Tudor Vermin Acts of 1532 and 1566, 69

Uganda, 7, 10, 25, 127–143, 170–86, 175f9.1, 198–99
Uganda National Council for Science and Technology (UNCST), 133, 174
Uganda Wildlife Authority (UWA), 131, 133, 142, 174
UK Wildlife Trusts, 73, 73f4.3, 87nn5–6
United States, 8, 16, 107–23, 124n1, 124n3, 162, 198
urban areas, 8, 52, 54, 116f6.3, 119
urbanization, 114, 115, 119, 194
urban wildlife, 82

218 Index

values, 6, 8, 15, 18, 101, 108–9, 128, 149, 196
  animal species, 21–22, 119
  competing values, 1, 3, 16, 155
  Wildlife Value Orientations. *See* Wildlife Value Orientations (WVO)

'Weapons of the Weak' (Scott), 7–8, 161–62
  conflict narratives as, 7, 26
White, T.H., 71–72
  The Sword in the Stone, 72
'wicked problems', 1–2, 155
  elephants as a 'wicked problem', 26, 29
wilderness, 5, 54, 59, 60, 63n1, 200
wildlife agencies, 11, 112, 113, 117, 123
wildlife foraging behaviour and crop protection, 170–73
wildlife management, 55–61, 56f3.2, 62, 110, 129, 148, 152, 186, 196
  conflict, and, 148, 152
  hunting, and, 8, 112
  Wildlife Value Orientations, and, 110, 112–13, 115, 117
  wolves, 60–61, 117–21
wildlife trade and 'conflict', 16, 22
  ivory, 16, 19, 25, 26, 29, 160
  rhinoceros horn, 16
  tiger bone, 16
Wildlife Value Orientations (WVOs), 8, 107–24, 142, 143
  attitudes and behaviour towards wildlife, 108–109, 111, 112, 117, 121–23
  cross-cultural perspective, 122
  Distanced individuals, 110, 112, 113t6.2
  domination orientation, 8, 108, 109–110, 111t6.1, 112, 114–17, 121
  education, 110, 11t6.1, 113, 114, 115, 116f6.2
  informing conflict management, 108, 110–14, 117–21
  modernization and shifting WVOs, 114–17
  mutualism orientation, 8, 108, 109–110, 111t6.1, 112, 114–17, 121
  Mutualists, 8, 110, 112, 113t6.2, 115, 115f6.1, 116f6.2, 116f6.3, 118, 118f6.4, 119
  Pluralists, 110, 112, 113t6.2
  Utilitarians, 8, 110, 112, 113, 113t6.2, 118, 118f.6.4, 119, 122
wild pig (*Potamochoerus* sp.), 9, 129, 136, 138, 143, 199
  bushmeat, 142
  crop damage by, 131, 176, 177
  legalized hunting of, 9, 131
  ranking of 'pest' species, 9, 137–40, 142
witchcraft, 83, 99, 160
wolves (*Canis lupus*), 5, 8, 49–63
  anti-wolf attitudes, 8, 52, 56
  conflict mitigation, 50, 55–62
  culling, 50, 56, 162
  encounters with, 8
  fences, 50, 57–60
  hunting dogs, 50, 52
  licenced hunting, 56, 58
  monitoring, 60–61
  population recovery in Scandinavia, 50–51
  pro-wolf attitudes, 5, 54, 56, 60–61, 119
  re-establishment of population, Washington state, US, 119
  sociological studies of wolf conflicts, 51–55
  symbolism, 5, 50, 54
  threat to human safety, 119
  Wildlife Value Orientations, 5, 117–21
  zoning and regional management, 55–57
working class, 50, 52, 59, 75
World Parks Congress, 4, 149

zoonotic disease transmission, 65